Care of Favorite Dolls

Antique Bisque Conservation

by Mary Caruso

Published by
Hobby House Press

Hobby House Press, Inc.
Grantsville, MD 21536

Dedication

To my Father, who, though he also had six sons to raise, spent great amounts of time and patience, showing me, his only daughter, the quiet dignity and rewards of working with my hands.

Thanks, Papa.

Additional copies of this book may be purchased at $24.95 (plus postage and handling) from
HOBBY HOUSE PRESS, INC.
1 Corporate Drive
Grantsville, Maryland 21536
1-800-554-1447
or from your favorite bookstore or dealer.

©1999 by Mary Caruso

All rights reserved. No part of this book may be reproduced or utilized in any form or by any means, electronic or mechanical, including photocopying, recording, or by any information storage and retrieval system, without permission in writing from the publisher. Inquiries should be addressed to Hobby House Press, Inc., 1 Corporate Drive, Grantsville, Maryland 21536.

Printed in the United States of America by Image Graphics Inc., Paducah KY

ISBN: 0-87588-545-4

Table of Contents

Chapter 1
Identification and Description 5

Chapter 2
Features of Antique and Reproduction Dolls 12

Chapter 3
Careful Handling of Antique Dolls 21

Chapter 4
Doll Repair 26

Chapter 5
Eye Repair 65

Chapter 6
Wig Care 89

Chapter 7
Dressing Your Dolls 113

Index 141

Biographical Information 143

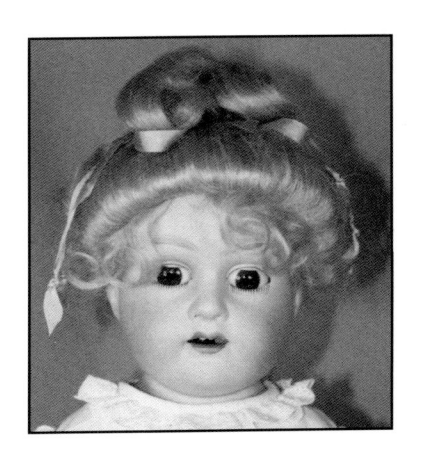

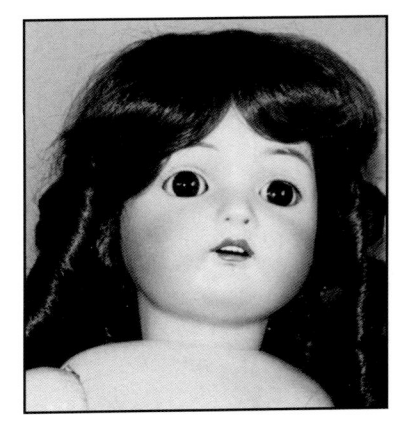

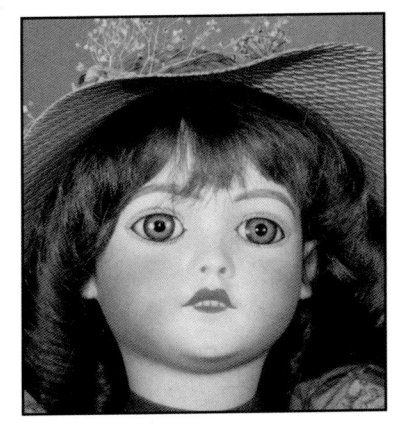

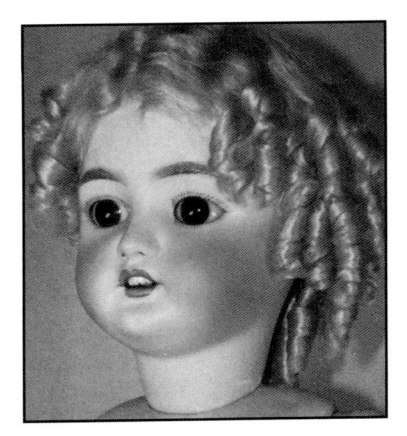

With Best Intentions

My purpose in writing this volume is to address the many questions about the care and costuming of dolls that my acquaintances and friends bring to me. Discussing every type of doll would require several volumes, a challenge which remains, pleasantly, in the future.

It seems logical to progress in discussions from older to newer dolls; hence this first volume will deal with one category of antique dolls, the bisque head child dolls of French and German origin.

Future volumes will cover antique china heads and parian dolls, antique bisque head baby dolls, composition dolls of the 1930s and 1940s, and hard plastic dolls from the late 1940s to early 1950s.

Since reproductions of the antique German and French dolls require the same kinds of techniques in stringing, wigging, and costuming as do the antique versions of these dolls, they are also included in our discussions.

Dolls and doll lovers have brought me great pleasure over the years, and it is hoped that in presenting this volume, I may share some of this joy with you.

So come have fun with us.

Taking Care

This book is about exactly that: taking care of your doll. Taking care means doing the right thing for the doll, and my pledge has always been to do what is in the doll's best interest...what is most "right" for the doll, given the limits of time, material availability, and of course, money. Hence, the often golden decision of doing less rather than more.

Often, repairs that are done to dolls do not enhance them in any way. It is up to the doll's current owner to exercise good custodial decision-making about repair concerns. After all, the dolls will outlive the owners, and need to be handed to the next care-takers, the better for having been in our hands and hearts.

So first, a few introductory comments about what is best. Basically, it is important as in the works of good care-givers everywhere, to "do no harm". This would mean no irreversible repairs or repairs of an incorrect nature. The doll owner who needs eyes in an antique doll and really can't afford proper glass eyes may exercise this principle well. Inexpensive plastic eyes may be used to keep the budget intact and allow the doll to be displayed and enjoyed. Simply, the eyes must be set in, in a reversible manner, so that "someday" they can be replaced with proper glass eyes. This would be an example of good guardianship.

This book will strive to remedy the many common problems of doll care. While acknowledging that collectors would all like antique dolls in pristine condition, this is often not the situation we are handed. Taking a realistic approach to repair while "doing no harm" to the dolls allows them to be restored to a realistic standard, for our times, and enables them to be enjoyed. Too often, wonderful old dolls lie poorly stored because the owners assume that restoring them to displayable condition will be prohibitively expensive.

So it is in this spirit that this book is presented: to recognize the best methods and materials, but to acknowledge the reality of the repair situation and to offer acceptable, less costly alternative methods, as well.

It is hoped it will be enjoyed by all.

Chapter 1

Identification and Description

Antique dolls are generally considered to be those made with bisque, parian, or china heads, most of them in Germany and France, and usually before the middle 1920s.

This volume will deal with antique dolls which have bisque heads. Other antique dolls, such as china heads, parians, and papier mâchés, will be described to make the distinction. Since their care and clothing sizes and styles are so different, they will be the subject of another volume.

Describing Antique Doll Head Materials

A definition of these terms is in order. Antique doll heads made of porcelain are described by the types of finishes given to the porcelain. Bisque, china, and parian heads were all made of the same type of clay. It was fired to a temperature of approximately 2400 degrees Fahrenheit; at this stage of manufacture the heads were white in color. Parian heads were left this white, matte finish, and the features were applied, such as cheek coloring, eyelashes, eyebrows, and lips; additional firings were done to render these colors permanent.

Bisque heads would have been tinted to a flesh shade and re-fired at a lower temperature. Then the features were added, with additional firings, again of a lower temperature, so as not to burn off the paints entirely. To make a china head, the procedure varied again. To the white head, after the first hard firing, a glazing material was added. It was then fired again at a different temperature, lower than the high, hard firing done first, but higher than the temperature for the firing of the features. When observing a china head, it may seem as though the glaze was done last. This was not the case, as the higher temperature required to melt the glazing material into the glossy finish would have burned off the paint used for the features. While these details of porcelain making may not interest all, it does help to know the correct terminology when discussing dolls. A knowledge of how finishes are applied also aids one in the critical inspection of a doll.

Dolls with bisque (porcelain with a matte finish and a flesh coloration) heads usually had "composition", leather, or leather-like "kidoline" bodies.

The ball-jointed "composition" bodies were actually made of a combination of materials. The torso was made of cardboard which was pressed, wet, into the two piece torso mold, front and back, trimmed and allowed to dry. It was then stapled and/or glued together at the seams, fitted with neck, arm and leg "sockets", finished and painted. Wood was used for lower arms, and the "ball" joints. The legs and upper arms were usually cardboard or a "composite" (probably mâché with some plaster or flour). The wood pieces were lathe turned; the other pieces were made in molds.

Making and painting all the pieces, finishing, painting, and assembling them into a body (called "stringing") was a time consuming task.

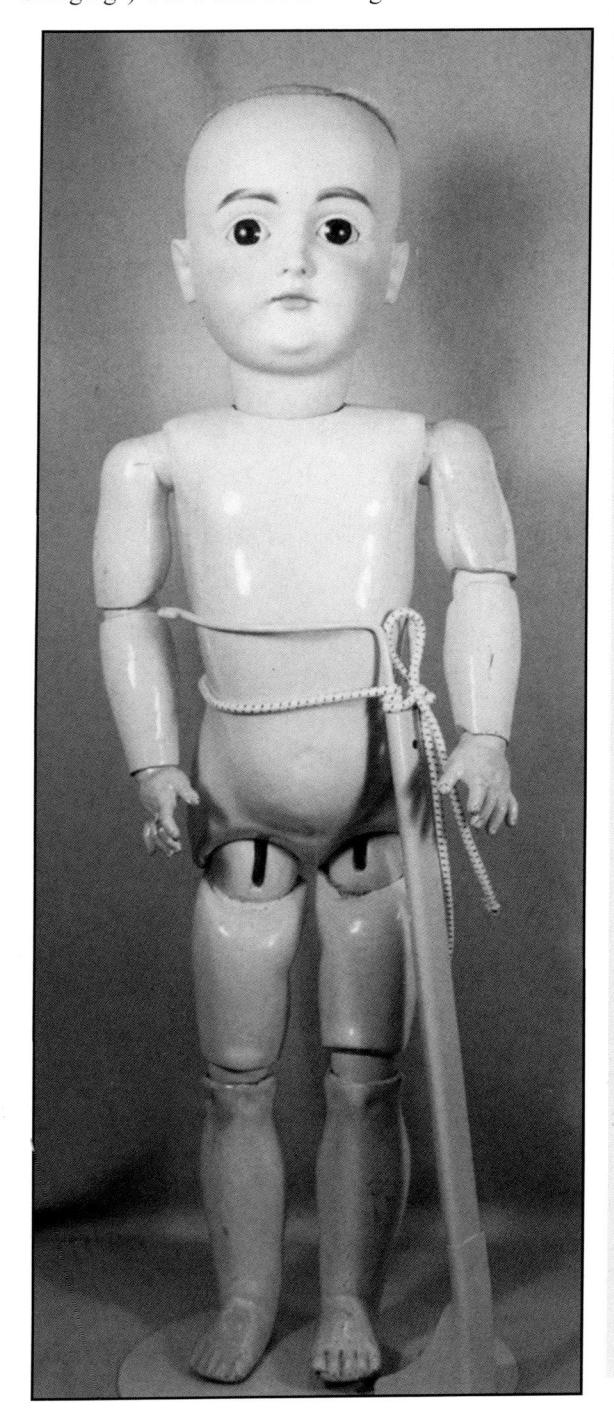

Front view of an antique closed mouth Kestner #128, original brown glass sleep eyes, original ball jointed body.

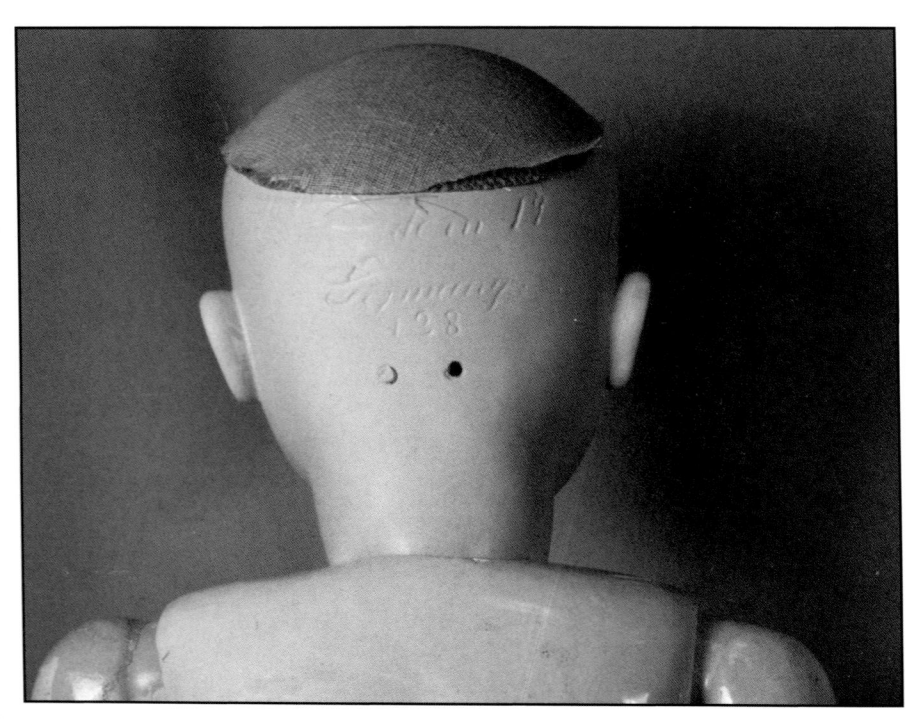

Close-up and back views of an antique closed mouth Kestner #128, original brown glass sleep eyes, original ball jointed body.

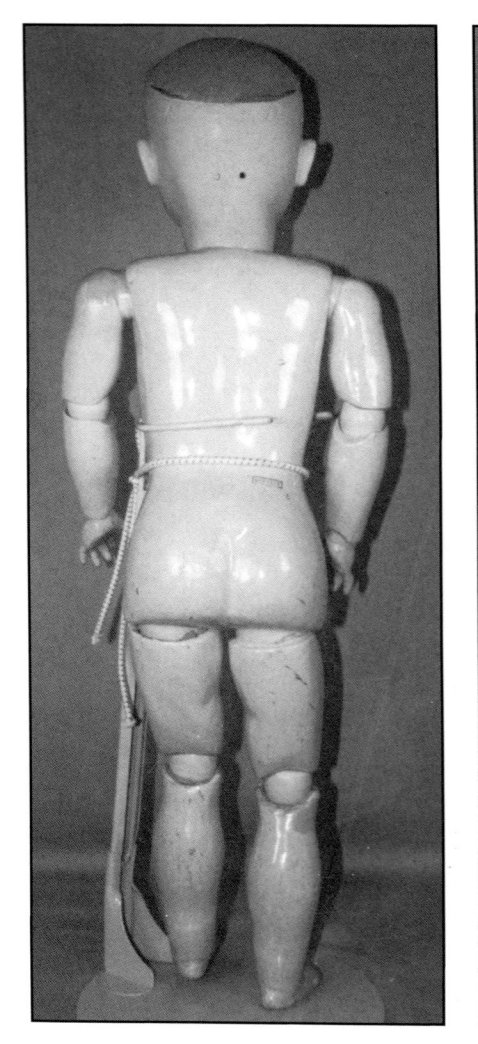

Antique Kämmer and Reinhardt doll, ball jointed body, front and back views.

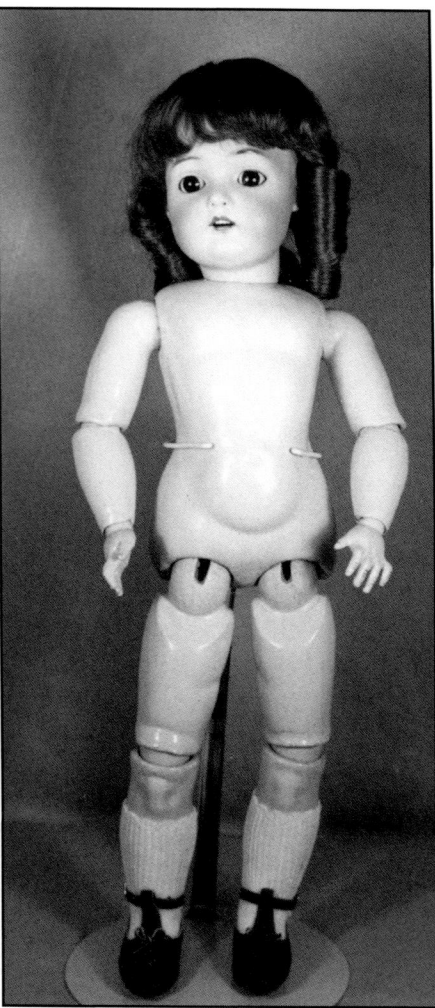

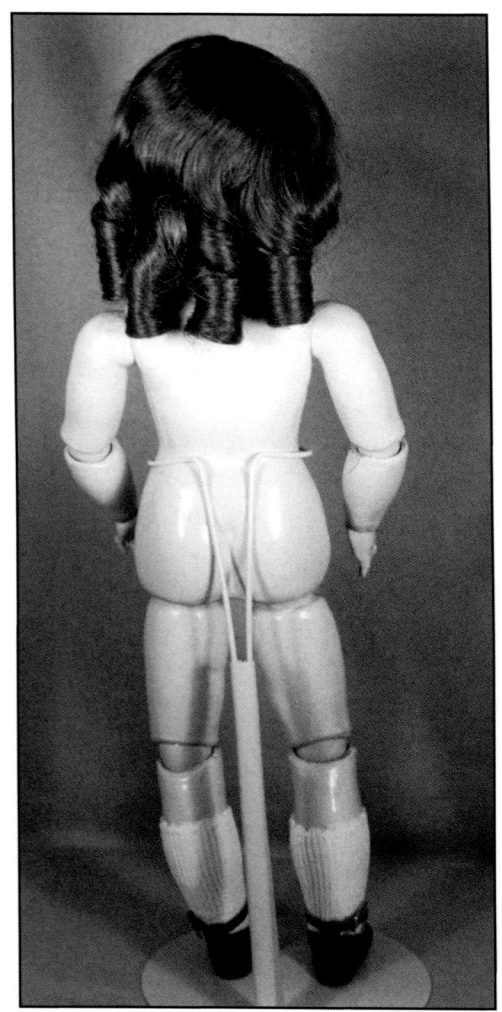

Later (from the 1910s to 1920s) socket head dolls in smaller sizes were sometimes done on what is called a five piece "crude" body. Due to decaying economic conditions in war-ravaged Germany, dollmaking firms had to make do with what scant materials could be obtained. Many smaller dolls from this period are notably lacking in quality, both in the face painting and in the body work.

Dolls with china or parian heads usually had cloth or leather bodies. There are probably a few rare exceptions to these statements "out there" in someone's attic, so it's safer to say "usually" and "generally". These descriptions will be more than adequate and useful for the majority of these dolls.

Most, but not all, antique bisque dolls are "marked". This means there was, in the mold, lettering and/or numbering to identify the maker, model, and country of origin. With experience, one acquires a feel for painting styles, and can usually distinguish between French and German work when the dolls are unmarked. These marks will be on the back of the head, sometimes under a wig, or on the back of a shoulderplate. China and parian heads are often unmarked.

A lovely antique Heinrich Handwerck doll, with original mohair wig, circa 1890.

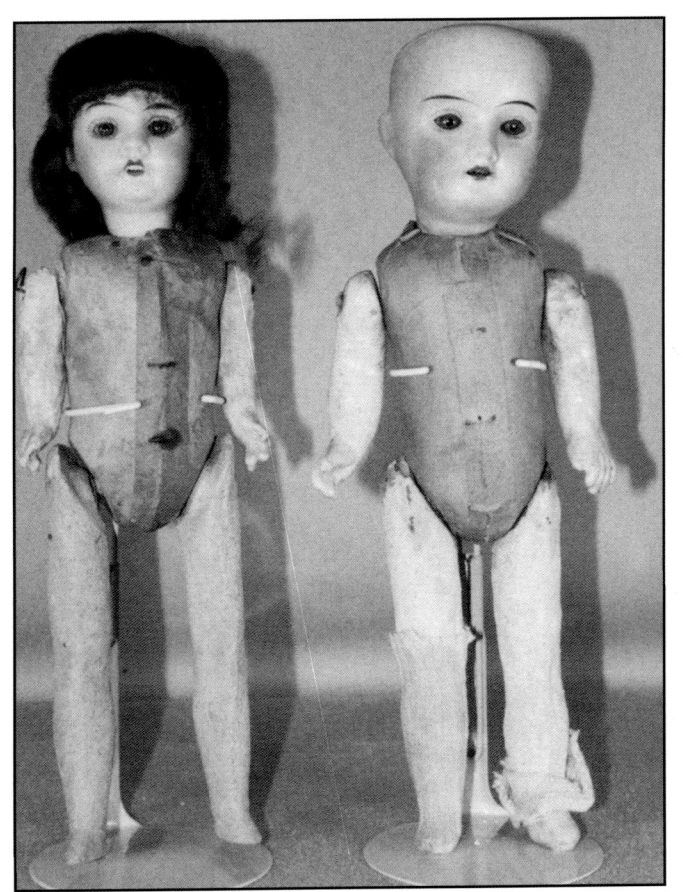

Small antique German bisques on 5-pc "crude" bodies, circa 1910-1920, by Recknagel.

Left, antique German parian, circa 1870-1880, ornamented hair, painted eyes, pierced ears. *Collection of Elizabeth Olson*; *right*, antique German china head, circa 1880, exposed ears, well-detailed blonde hair.

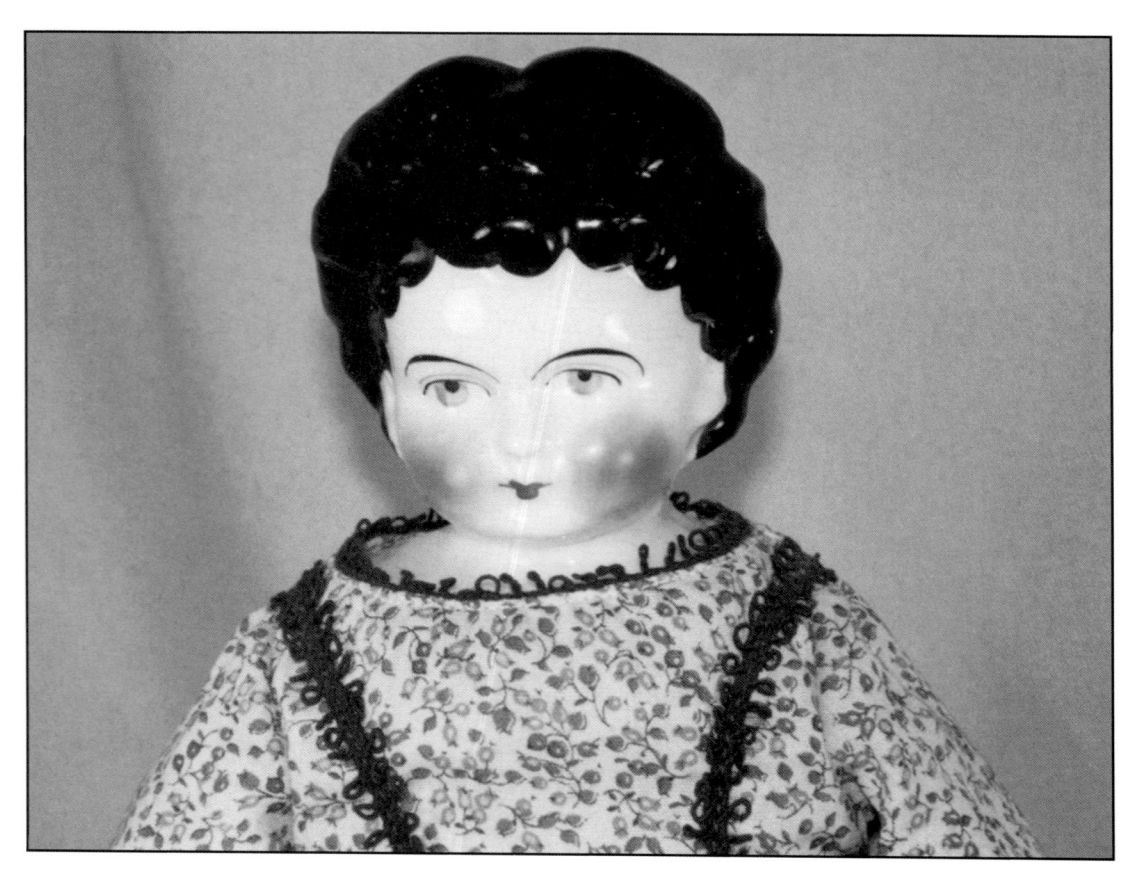

Antique German china head, circa 1890, highly colored rosy cheeks.

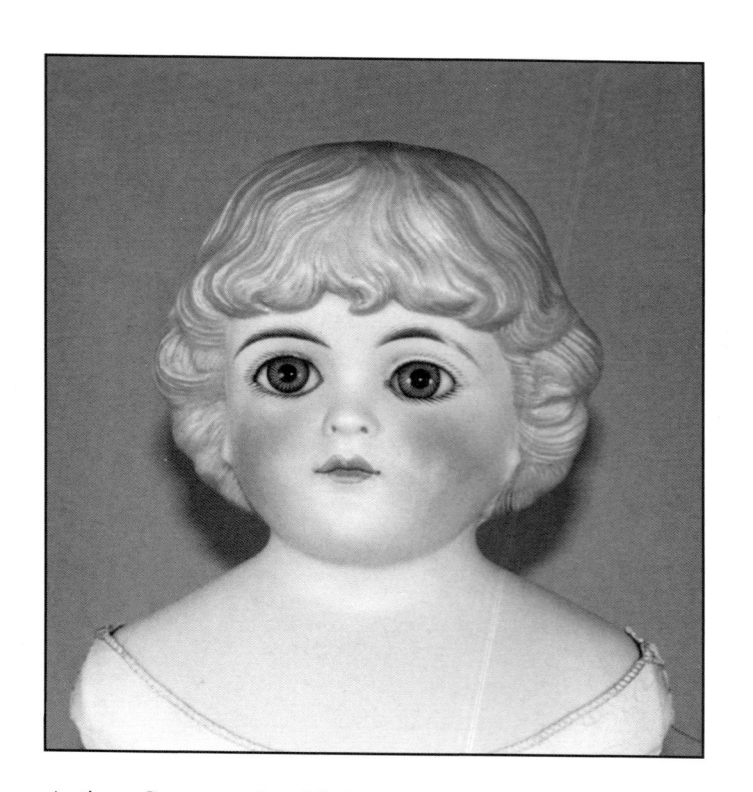

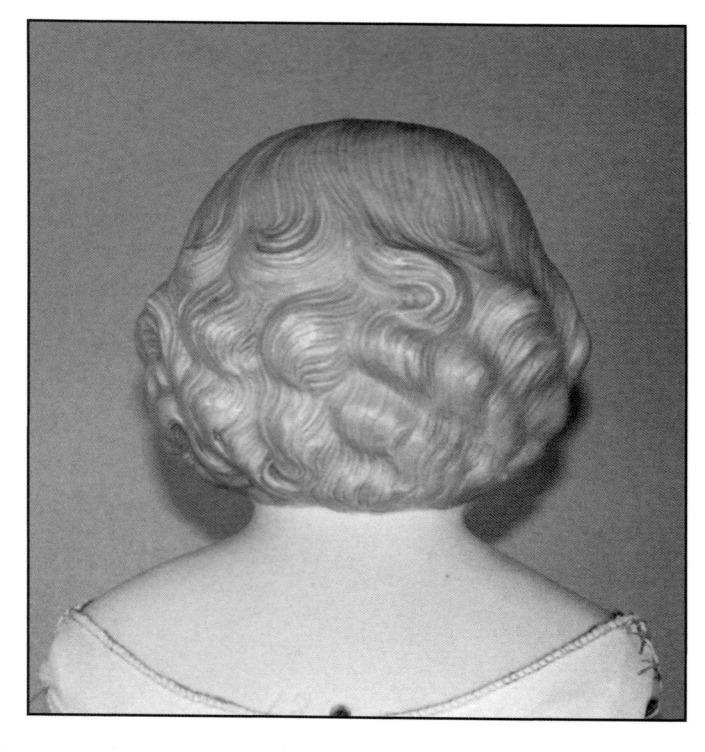

Antique German painted-hair bisque head on original kid body. *Collection of Nan Sheehan*. The only visible mark is size number "12". This is a seldom-seen type of bisque doll, with beautiful over-sized original set glass eyes, wonderful coloring and detail in the hairstyle. She's a gem of a doll!

Specifics in Describing a Doll

It's very helpful to get in the habit of mentally listing doll specifications when looking at dolls for possible purchase, or when studying them. Practicing this mental exercise will make you much more observant of detail.

More serious reasons also exist for learning how to pinpoint the specifics of an individual doll. What points would need to be mentioned to give a specific description of one, for example, for an insurance statement or a will?

To describe a doll accurately, begin with the head. Material, size, marks, type and color of eye, type, color, and style of hair, and any defects or breakage would all be included.

The head description might read: 12" circumference, antique German bisque socket head, marked "made in Germany, A. 390 M.", original blue glass sleep eyes, blonde mohair wig, center part, long curls; slight scuff to right cheek. All these items mentioned help specify the doll's origin and any distinguishing features.

To describe the body, one would add: "on a 15-piece ball-jointed body, total height of doll 25 inches, missing fourth finger left hand." Of course, a description of clothing could also be added, and each doll could be given a collection number, on a small paper tag attached to the doll.

With this type of description, an accurate appraisal could be written. Anyone having an antique doll in the household would be wise to have this type of documentation written up. For insurance purposes, an advance copy of this type of description in the hands of your insurance agent would make the processing of a claim much simpler. Attach a photograph of the doll and a notation of the doll's value, giving the price guide and page number of the listing.

Often, doll professionals are asked by an insurance company to evaluate a claim when a loss has occurred. If all the agent has for information is "the doll in the blue dress", it makes it very difficult for any appraiser to suggest an appropriate settlement amount.

This type of documentation of significant dolls should also be attached to the collector's will and any other legal documents dealing with the disbursement of his or her estate. This accurate, advance reporting of each item in the collection, with specific details, makes the task of identifying each item much easier for heirs and their legal representatives.

Comparing Antique and Antique Replica Dolls

Whenever and wherever doll owners gather to talk about dolls, questions arise about identification and verification of the age of their dolls. It seems wise to address, in broad terms at least, the features of antique dolls and of the replicas of antique dolls, which are enjoying an enormously expanding market.

Whether you prefer old or new "antique" dolls, it is hoped you will enjoy, use, and benefit from this book. Many of the areas addressed apply to both types of these dolls. Many collectors of antique dolls will purchase the replicas of the very rare dolls, as they feel they will never otherwise own these "faces". Others, who aspire to create exquisite reproductions, can only do so when they have learned just how the old dolls were made, and should look. These creators of the "new antiques" find their art and craft grounded in their study of the old dolls. So, antique or replica lovers, all are welcome.

It would be well to begin by dispelling any confusion of terms. Antique dolls are just that, dolls of approximately 100 years in age. For our purposes, we will soften the age requirement a bit and consider any of the bisques produced into the 1920s as antique dolls. Antique reproductions are COPIES of these dolls. We won't call them fakes, as they are quite real, just not old.

We are not addressing in this chapter the enormous industry in commercially produced porcelain dolls, mostly imported from the Asian shelf. These, for the most part, cannot be considered antique reproductions, as there is little, if any, resemblance to the original dolls.

Some faces will be very recognizable, however, as they are being "done to death", sometimes quite literally. This intriguing subject area will be covered in a separate section, "Adoptables", as some are quite irresistible, and availability is so high that it is unlikely any of us will be able to escape this earth without having owned a few of these dolls!.

Reproductions, or "copies" of antique dolls abound. The great expansion of the doll mold industry, beginning in the 1960s and exploding in the 1980s, placed molds of antique reproduction doll heads in the hands of many "dollmakers".

Since these molds were made from actual antique heads, they bear all the same markings as the original heads. This can lead to great confusion for the general public. The novice doll buyer is wise to seek the advice of an experienced doll person before buying what they believe is an antique doll, both when the price is remarkably low, and even when it isn't.

Since most reproductions are done by craftsmen/women without the proper training in painting techniques, these heads are usually easily identifiable. There are many dollmakers, however, who are expert in producing the painting techniques required in copying these old dolls. The great majority of them are very honest and tag and sign their doll heads as reproductions. The problem arises when these dolls are on the secondary market and the potential buyer cannot consult the doll maker. If inherited or found at a yard sale, these reproduction dolls often lead to confusion for seller and buyer.

Reproduction dolls are often the choice of purchase for many doll buyers who admire the antique styles. Most of us will never find nor afford to buy a Bru, for example, and are happy with a well-executed reproduction for a few hundred dollars.

One should distinguish, however, between a well-executed example done by someone with outstanding painting skills, and all china-fired, on the proper body, as opposed to commercially produced dolls which are done by the millions, not to the proper standards, not on the proper body, and with factory spray-painted features rather than the above mentioned china-fired faces. These commercial dolls are done in such enormous quantities that they will not be rare for several generations, if ever, and should not command prices that even approach a hundred dollars if the buyer is hoping to build investment.

Occasionally, a visitor lays out on the doll counter a Bru or Jumeau, which they have just purchased for a few dollars at a flea market or yard sale. Confident that the seller was uninformed about dolls, they believe that they have made a "find", and anxiously await an appraisal in the thousands or tens of thousands of dollars.

Imagine their disappointment when we tell them that the doll is "new", homemade, and worth whatever amount of pleasure it gives, but a modest sum in dollars. Reaction is varied. Some take it philosophically for the lesson that it is when we point out the various factors which tell the difference between antique and reproduction. A few others let us know they are sure we are mistaken.

Doll "Stories"

Otherwise titled "Guess what my doll is?"

This game is especially "fun" when it is enacted over the phone!

For this version, our scenario is face to face, and I can see the doll in question!

A common and humorous (to me, at least) assumption by some is that a doll is as old as its owner. The dialogue is generally in the vein of someone wanting a doll appraised that "Aunt Mildred, who is 85 now" gave them. A few questions elicit the information that the head is "glass", meaning, usually, porcelain. It is a dull white finish. The arms and legs are the same. The body is cloth, the eyes painted. Of course, this could be an antique bisque head on a replacement cloth body, or a valuable parian. When the doll is finally made available for a visible inspection, we find that it is an imported product of Taiwan, made in vast quantities in the early 1980s and advertised on the back of every Sunday magazine section in the country for the sum of $4.95. Alas, the heir is sure that our appraisal is incorrect, and Aunt Mildred really played with this doll when she was a little girl.

Unfortunately, a few unscrupulous dealers will add to the problem of doll identification by weaving untruthful, but charming, tales of how the doll came over from Europe in the arms of her little owner. While this is uncommon, I have been subjected to this type of "information" myself a few times. Yet it is fairly easy to learn many of the points which distinguish antique bisque dolls from new porcelains, and to distinguish good quality features from lesser quality ones. One rule (and there will be a few!) to maintain is to always be curious, even skeptical, when studying a doll; another rule is to be even more skeptical when the price seems too good to pass up.

Distinguishing between Antique and Antique Replica Porcelains

This section is intended to help the novice doll enthusiast when he or she is considering a doll purchase Those readers who are "veterans of the hunt" may want to pass over this section, or skim it for a few tidbits of humor, exalting that they have passed through this stage no worse for the wear.

In learning about dolls, no lecture, article, or book will replace actual experience. Experience will be your best teacher. Careful observation of a great number of dolls will provide much knowledge, but of course, for most of us this will take some time.

How to Learn about Dolls

While much can be learned by talking to those with a depth of experience, your best teacher is the doll itself. If you are in an area with good availability of dolls, many doll shows and events similar, you will be amazed at how quickly you will learn.

Unfortunately, to learn the most, the dolls need to be undressed, and this is often not possible at doll shows or when viewing others' collections!

Anyone who works with dolls, the so-called "Doll Lady" (or "Doll Guy"!) is always asked how they learned about dolls. Did they take a class? This is such a difficult question for me to answer, I always want to reply that the dolls taught me!

Occasionally someone will ask, "Do you doll people really talk to your dolls? We've heard that some of you do this!"

And we reply, "Oh, no, that's just a rumor! "And turning, ask one of our dolls, "Isn't that right, honey?"

There really is no substitute for acquiring a few dolls, slowly and cautiously, and learning from them what they are made of and how they are put together. In time, you will acquire a good "feel" for when a doll and her accouterments are "right" and original, and when they are not. With time and a sharp eye, it will seem as though the doll is actually telling you these things.

Antique or New?

For those who would want to learn, the following are some features to observe:

Antique doll heads were poured with slip which was white when fired. It was then tinted and fired again to produce the flesh coloring of bisque heads.

Parians were left white, and facial colors were completed and fired. They are untinted and unglazed (matte finish).

China heads were done usually in white, as the parians, glazed, and then facial features were applied, so they have a white glossy finish. A few have a pale pink tint, and are called "pink lustre". They also have a glazed finish.

Reproduction dolls are usually (not always) poured in a flesh-tinted slip (slip is a term for clay in a thick liquid form which can be poured into molds. Some old dolls were made of clay which was rolled out like cookie dough and then pressed into the molds). This "tinted slip" saves the crafter the step of tinting (and firing again) the bisque. Checking inside the head will reveal which color of slip was used, as originally tint would only have been applied to the outside. If the porcelain is flesh-tint all the way through, the head is "new". If the head is white inside, tinted outside, look for other signs of age. With experience you will get a "feel" for when porcelain is new and when it is not.

What Dolls to Buy? Old or New?

I have happily noticed how the buying habits of some of my friends changed, slowly, after they began to "hang out" at the doll shop and started going to every show and auction that they could for purposes of drenching themselves in the doll world. Soon, they were talking of "upgrading their dolls," and "trading up" their collections. While I couldn't (and didn't wish to) take credit for their new-found expertise, I am happy for us all.

Actually, most of us do this, just as we try to improve our minds or our wardrobes or the quality of our lives. So, shop on, doll people, there are many more dolls out there for you to find and love.

Reproduction dolls, as their ancestors, bring far more joy than vexation. Many crafters relish the various challenges of dollmaking, and presumably would welcome suggestions in their care and costuming, just as do the owners of their dolls' antique counterparts. Some doll lovers enjoy pairing "sisters" of a given doll mold, one antique and one a reproduction.

May collectors of both types of these dolls enjoy and benefit from these chapters.

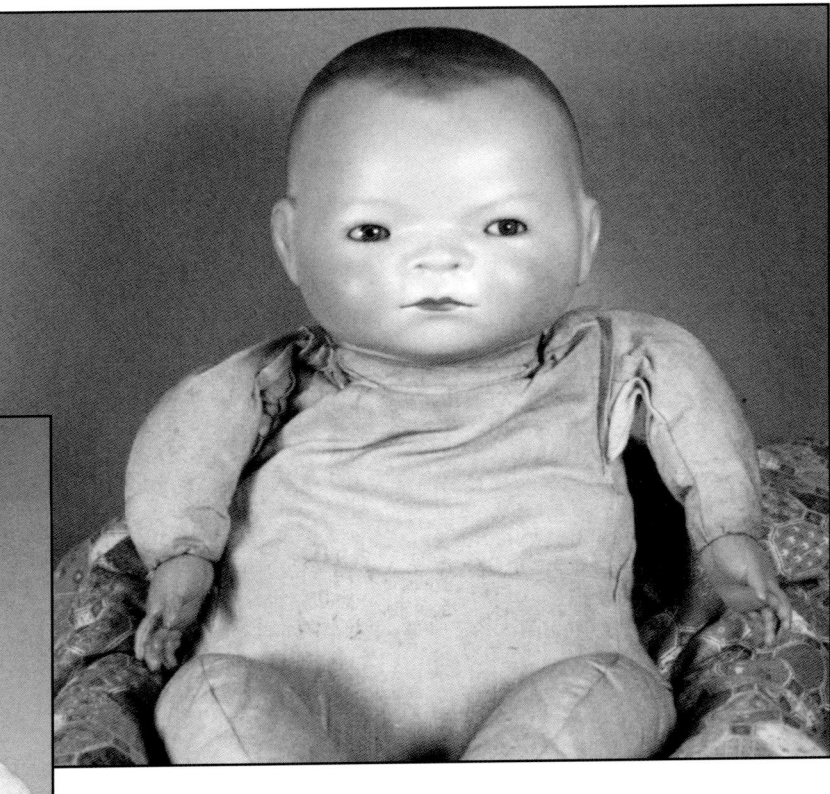

Antique *Bye-lo*, front and back views, original body, note celluloid hands, distinct mark "Grace Storey Putnam". Germany circa 1920s.

Chapter 2

Features of Antique and Reproduction Dolls

This chapter is intended to help you gain expertise in evaluating a doll, whether antique or new. All too often, a doll is brought to me by someone who has just purchased it. Before a word is spoken, I know he or she wants me to tell them everything I know about this doll, and of course, whether it was a "good buy".

I walk this minefield carefully, mentioning the quality features and trying to give them "good news" about their new doll. It's often disappointing, however, to have to reveal that the eyes are poorly fitted, the head is cracked, the body is "put-together" or the wrong type or size.

We have all purchased dolls that had more or less these problems, perhaps because we wanted to save them, or restore them. It's discomforting, however, when someone purchases a doll thinking that it is in good condition and is later disappointed to learn of all its detracting features.

So it is my goal here to help you to sharpen your eye in the critical observation of the various features of dolls, antique and new. Then if you buy a little "basket case" because you love her, well, we have all done that.

We will cover several areas of study, as follows:

Eye Characteristics

To accurately evaluate a doll, being able to distinguish new eyes from antique eyes, and glass from plastic eyes, is essential. When you are buying a new doll, you also expect the eyes to be new. However, there is a wide range of quality in new doll eyes, and several kinds of materials. Are they glass or plastic? Should they be" sleep" eyes? Are those better? Are set eyes less desirable? Are painted eyes?

The price for a doll with two dollar plastic eyes should not be as high as for a doll with high quality acrylic eyes. If you are buying an antique doll which does not feature painted eyes, such as many china-heads and parians, the eyes should be original and of the correct type.

Painted eyes, if done artistically, would actually make the doll more an individual work of art than if mass produced eyes were inserted.

The Correct Eye Type

In certain old dolls, "sleep" eyes would be expected. For example, antique dolls of the "dolly face" variety (Armand Marseille dolls of molds 370/390/Floradora/1894/3200, Simon Halbig, Heinrich Handwerck, Kestner...to name only a few) usually had "sleep" eyes, that is, eyes which operated on a rocker with a lead weight, which would close when the doll was laid horizontal.

It is common to find these old dolls with original eyes set in place, fixed. The eyes would commonly have dropped out during play (imagine these dolls were for children's play!!) and were possibly put back in by someone who didn't know how to reset them to operate, or the eyes had broken away from the rocker and chips of glass were missing.

Another common situation is the replacement of eyes in these old dolls with new glass or plastic eyes. The impact this has on desirability depends on how rare the doll is. A rare doll may be desirable to the buyer with replaced eyes, whereas a more common doll will surely be available in a more original state. Of course, price should always be relative to condition, degree of originality, and the relative ease with which the doll could be returned to its original state.

It is usually not possible to use an old pair of glass eyes on a rocker as replacement eyes for a doll which needs eyes. The reason for this is that eye rockers were built individually inside the doll's head and are uniquely fitted to that doll, both in "gaze" and in the distance between the eyes. Chapter Five, "Eye Repair" covers the various problems of eye repair.

Replacement sleep eyes can be achieved for antique or replica dolls. Using a pair of well matched, good quality hand-blown glass eyes, fitting into the particular head in question and building a rocker will give the desired result. The eyes will be new but executed in the old way.

French bebes and fashions and many German closed mouth dolls originally had glass paperweight eyes. Exquisite new glass paperweight eyes are available to the doll maker, but are not inexpensive. Expect these good quality glass paperweight eyes rather than low quality, inexpensive eyes, when you purchase a reproduction doll of this type.

New and Replacement Eyes

Recognizing that many doll collectors will want to enjoy a particular doll and not have the budget to put in the best, most appropriate type of eye, we do not scoff (at least, not terribly!) at using a lesser eye. Hopefully, this compromise will be temporary, and the eyes can be upgraded at a later time. There are some reasonable quality acrylic eyes on the market; technique is at least as important as the materials (in this situation), to yield an acceptable result. Using the proper, reversible techniques for installing set eyes allows the budget-constrained individual to enjoy his or her doll on display until a time when more funds are available.

Eye Quality

Whether old or new, doll eyes exist in a range of quality levels, from very poor to very nice. When you first begin to really study doll eyes, in dolls as well as (out of the doll) in your hand, you will see what many others overlook.

What should you look for in eye quality? Good color, adequate "threading" (the little white lines in the iris that make the eye look realistic), well-matched iris and pupil shapes, and of course well matched colors. It is not unusual to see less than well matched old pairs of eyes in old dolls. They were made as toys for children and usually had to be kept in a low price bracket, so eye quality was not an important consideration. There are also high quality old eyes in many old dolls, especially the "better names", such as Kämmer and Reinhardt, and Simon and Halbig, Handwerck, and Kestner, to name a few. In the French dolls, the antique paperweight eyes are usually gorgeous, with deep color and unbelievable detail in the threading.

When purchasing a doll with new eyes, expect a high quality, well matched pair, as these are readily available at reasonable prices. Expect glass eyes or a lower price for the doll if the eyes are plastic. Don't be fooled by names that begin with "glass" and end in "ic". Some lesser quality brands of eyes cost as much as a good glass pair would and don't begin to compare. Be particular about eyes, they are a very important feature of your doll.

New hand-blown glass eyes, used in "sister" dolls. The doll on the left is antique *Queen Louise*, Germany, circa 1900-1920. On right, a reproduction (new) *Queen Louise*, with the same quality glass eye.

New glass paperweight eyes, shown in reproduction (new) Bébé Louvre.

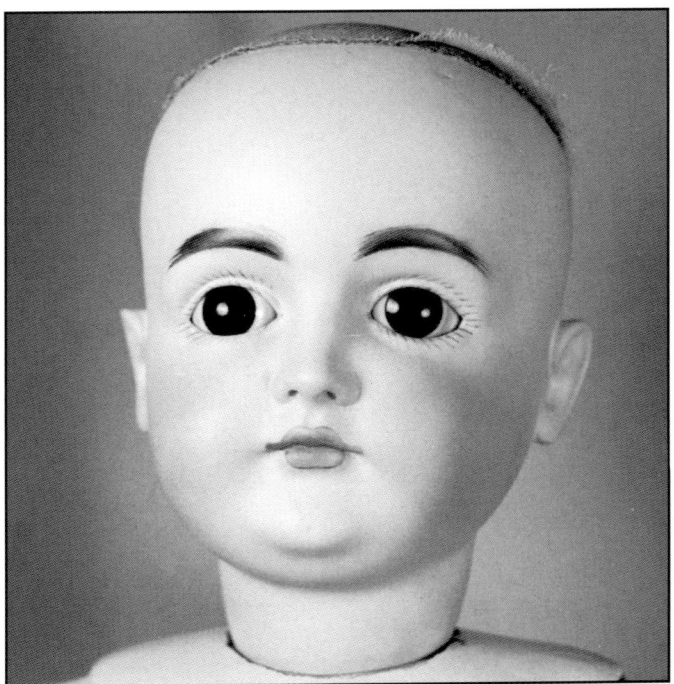

Original glass sleep eyes shown in antique Kestner #128, made in Germany, circa 1890-1900.

Evaluating Doll Heads for Quality

Let's use our critical skills for evaluating two doll heads. Of course, there is some handicap, as you are not able to inspect the dolls "in person", but can only observe the photographs (some extra information will be provided).

In the photos on this page, the subject is a C.M. Bergmann head, made by Simon and Halbig; a doll in the "medium range" of desirability, that is to say, a nice dolly face of between standard and good quality. What do you see? Responses would include: lovely, even facial coloring, delicate multi-stroke brows, nicely painted lips, all four teeth present, (no hairlines are present in this head, and original blue glass sleep eyes are functioning). The only detracting feature visible (look closely) in the front view photograph is the slight misalignment of the sleep eyes, a problem which can probably be corrected easily. In the rear view photograph, the white splotches are called "wig pulls", a term which means that someone got too rough in removing (pulling rather than soaking the wig off) the wig, and pulled some of the flesh coloration with it. A minor case of doll abuse, and unnecessary!

In the top photo on page 15, the subject is a large (36inch doll) Kämmer and Reinhardt head. One's observations would include: excellent smooth bisque, highly colored cheeks, lovely molded and lacquered multi-stroke brows, and original, well set glass sleep eyes with all the original lashes perfectly in place. Her original human hair wig has been washed and set; its brown shade matches her brows well and flatters her coloring.

Our third subject in the doll critique session is shown on the bottom of page 15. She is a mold attributed to Kestner, #196. Her facial tones are soft and even, with a pleasant little mouth and all four teeth. She has the remnants of the skin from her original fur eyebrows. Her lower painted lashes are the straight, rather than the curved variety. Some collectors prefer the straight up and down painted lashes, as they seem to indicate an earlier doll. But not all dolls, nor all Kestners, are created equally. Indeed, every head is the effort of different painting techniques and skill levels, by various persons. How could one price in a price guide be correct for all of them?

If this doll has a detracting feature, it is that her eyes, while original and functional, are not of the best quality, not well threaded, and without hair eyelashes, or upper painted eyelashes, they look, well, "bald", or a little surprised. Another aspect of the negative about her eyes is that they are too low in the eye opening, meaning that more of the blue iris shows at the top of the eye than at the bottom. This would not be noticed by the casual observer, but this is what critiquing a doll is all about, to really "look for problems".

While the head of an antique doll is usually "where the money is", meaning that the marks and quality of facial painting determine price, to a large extent, correct body type in good condition is also very important. A fabulous antique head on a decrepit or incorrect body will not be worth its usual value.

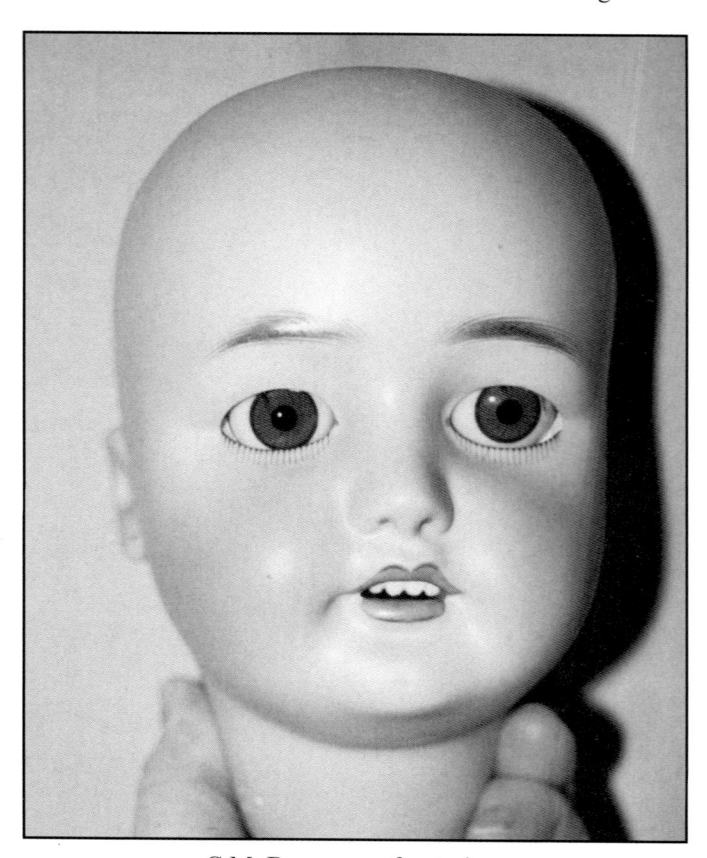

C.M. Bergmann, front view.

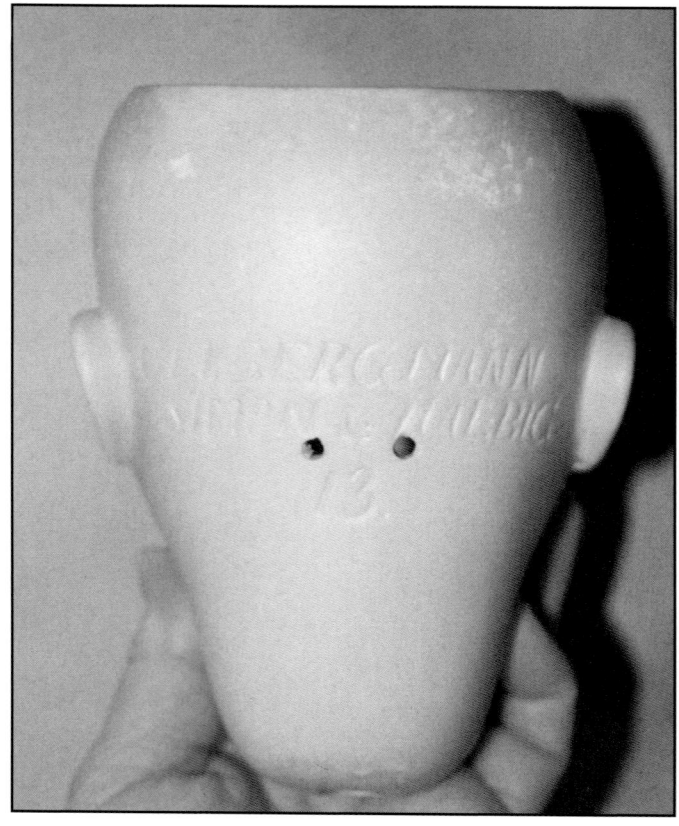

C.M. Bergmann, back view.

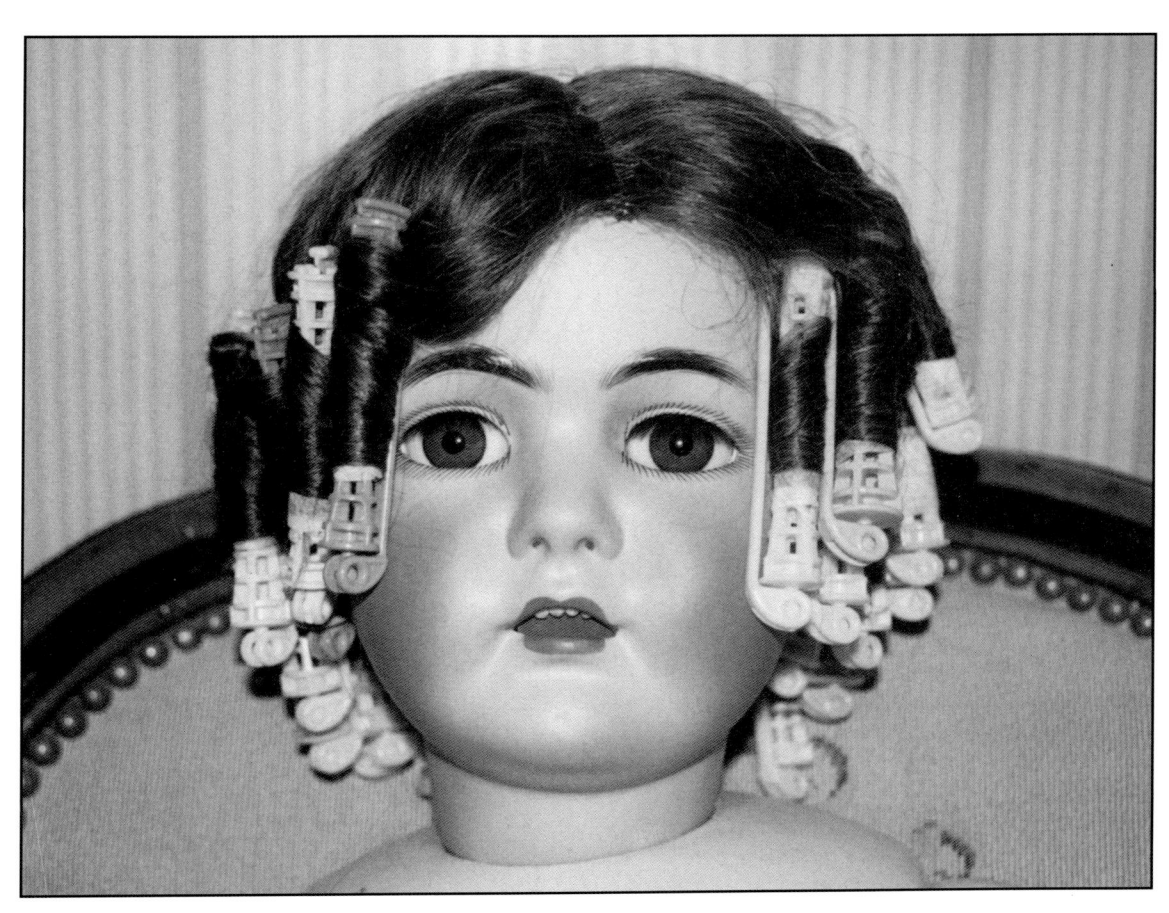

Kämmer and Reinhardt.

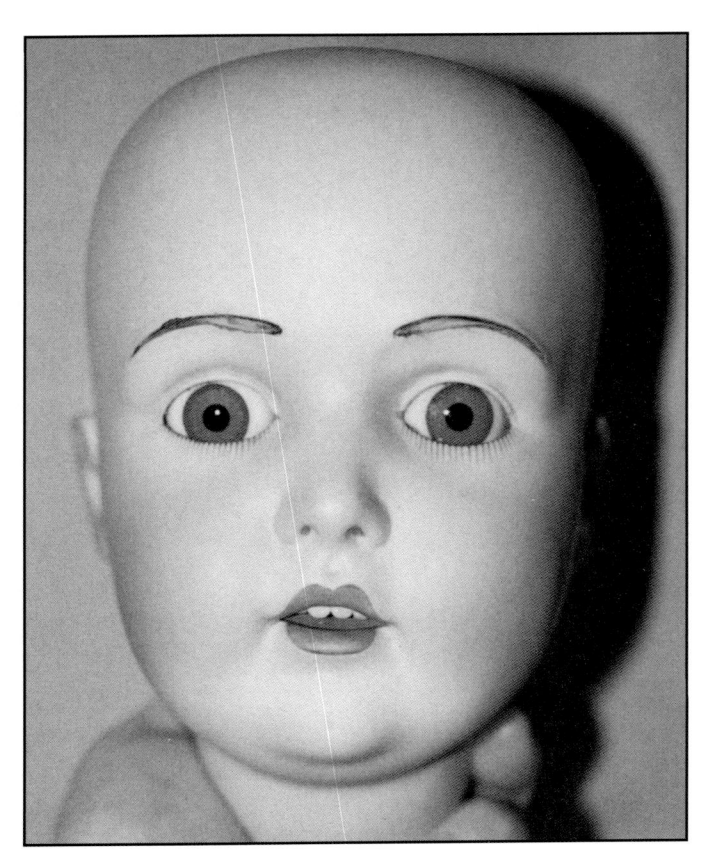

Kestner #196, front view.

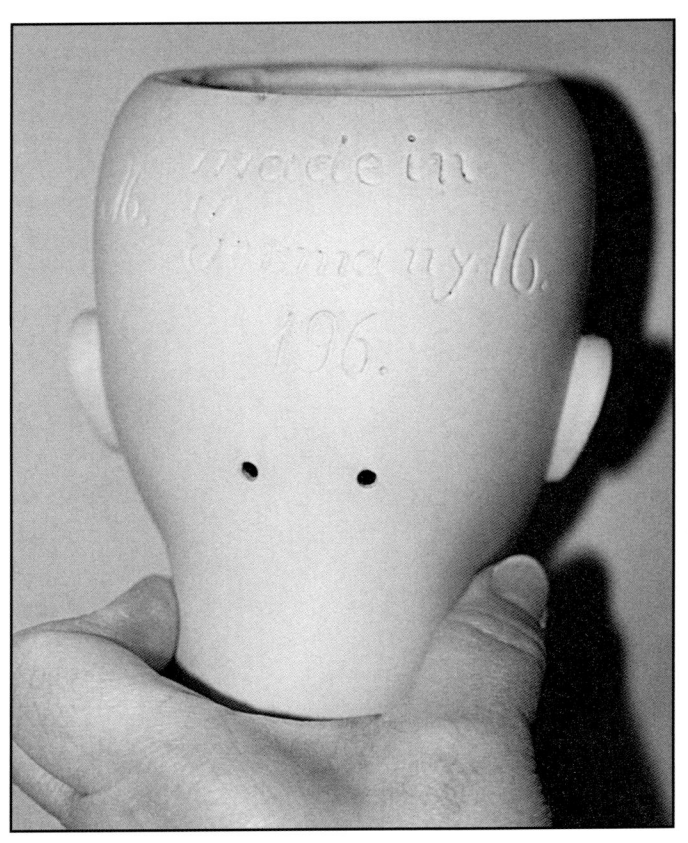

Kestner #196, back view.

15

Body Types

Reproduction dolls should be assembled on the same type of body as the antique version. It's disconcerting to see French bébés on a German-style ball-jointed body, or worse yet, on a porcelain body from the very available new molds. Dollmakers often "cut corners" in this area. Porcelain bodies are easier for them to reproduce (since they have the porcelain skills and equipment), and are much cheaper to acquire or make than a proper "composition" ball-jointed body. It is a prevailing myth, perpetuated by the mass advertising of porcelain, that porcelain is more desirable than other materials. We have given these heavy all porcelain dolls the dubious distinction of the title "cement babies", in honor of their excessive, unnecessary weight. Porcelain is often not the best choice of material for a doll body, particularly if it results in the wrong type of body, or if it is poorly executed.

Evaluating Dolls for Authenticity of Body

When considering the purchase of an antique doll, expect an appropriate antique body. "Appropriate" means the correct style, the correct period (age), and the correct size!

Often, dolls are "put together" with a head and body which are ill matched in size. Be observant about the doll's overall proportion, check that the head fits well into or onto the body. Look for signs of re-gluing on leather bodies, ill-fitting parts and mismatched colors of paint on ball jointed bodies.

Old Head, New Body

Too often, new ball-jointed bodies are used with old doll heads, and offered at doll sales and shows. Some publications even discuss how to make a new body look old by scuffing and dirtying the new body with newspaper. Most dealers are honest, and when asked, will tell you, or at least say they do not know if the body is old. The one comment which makes me wary of a dealer is, "Well, I don't know, it was like this when I got it". This may be the total truth, but it reveals a dealer who lacks the experience to assess the condition of the doll. How then, can he or she possibly price the doll appropriately?

So all of these concerns explain why it is essential to see a doll undressed before you purchase it, unless you have a long-standing relationship with your dealer and can relax in that trust.

Ball jointed bodies; left: new, German-style; center: new, French-style; right: antique German-style body, doll is *Pansy I*, circa 1920s, imported to America by Borgfeldt.

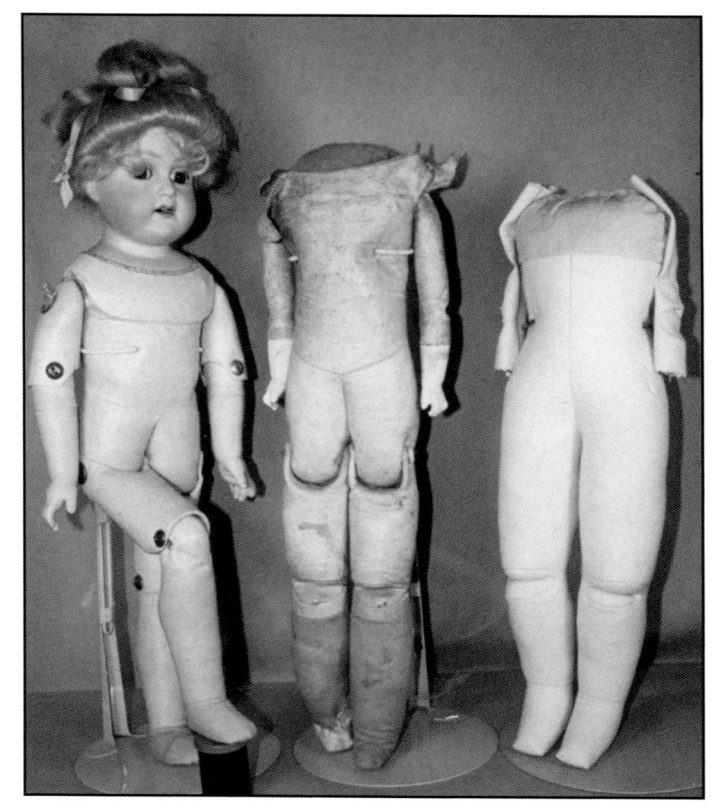

German body styles for shoulderplate heads. *Left:* Morimura Bros. (Japanese) doll of exceptional quality; center: typical kid-leather body; *right:* new vinyl "leather look" replacement body.

The "Put-Together" Body

An increasingly common problem in the antique doll market-place is the "put-together" body. Of course, if parts are missing, the doll will not be easy to sell. Some well-meaning folks will cast about for random old body parts, and assemble a doll, using these mismatched pieces along with what they already have. Of course, these "put-together" dolls are not wise purchases. Often, dolls are "put together" with a head and body which are ill matched in size. Be observant about the doll's overall proportion, and check that the head fits well into or onto the body. Look for signs of re-gluing on leather bodies and ill-fitting parts on ball jointed bodies.

Yet another layer of anguish to the put-together doll is the put-together body. It is possible to find dolls which have a variety of "borrowed" parts: arms from one source, legs from another, and so forth; all kinds of "organ donation" apparent. These dolls make the worst kind of purchase, because you are basically buying the head. It may cost you the purchase price amount again to acquire the proper body, if one is ever available to you.

There are occasional lucky breaks. I once purchased at auction a fabulous large Kestner child doll, marvelous in every detail...except that one lower leg was missing. The price, of course, was very low. I was amazed that I had in my accumulation of antique parts the perfect lower leg, matching in every detail of size, modeling, and color. As my daughter chimed, "Mom! You had that doll's leg in your basement!"

Usually it is almost impossible to acquire a proper match of an old part to an old body with acceptable results, and these "put-together" dolls will be obvious to the inquisitive eye. Time for another rule! You must ask for dolls to be undressed for inspection before you buy them!

I can hear some of you leaving now, to run up to your doll room, and check out bodies. I know I have made some of these "mistakes" and am living to more or less enjoy them anyway!

One example of a put-together is shown below. An antique "B-62" doll with at least four "borrowed" sets of parts. If you look closely, you'll notice that the upper arms, lower arms, hands, and lower legs do not match the torso, or each other, in color. This doll has the further problem of a crack in the back of the head, and paint missing due to sanding of the crack and wig "pulling". Although she has a sweet face, these other problems would render her a poor investment to make. Incidentally, we searched for a number of years to determine what kind of doll a "62" was. It's easy to stay humble, we finally learned that this number is the centimeter measure of the doll's height, about 25 inches!

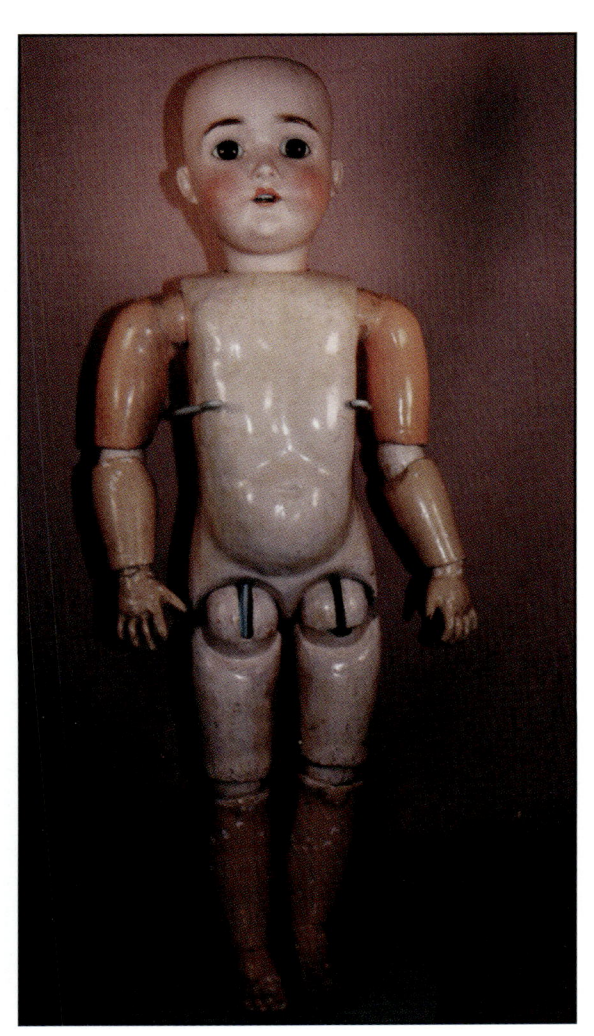

Antique B-62, front view.

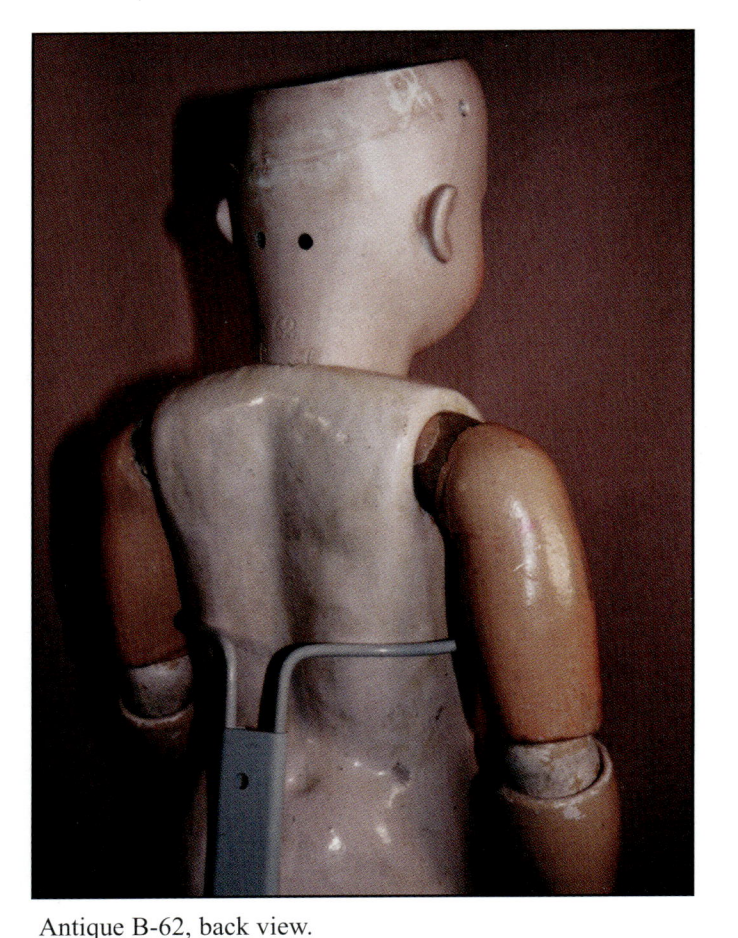

Antique B-62, back view.

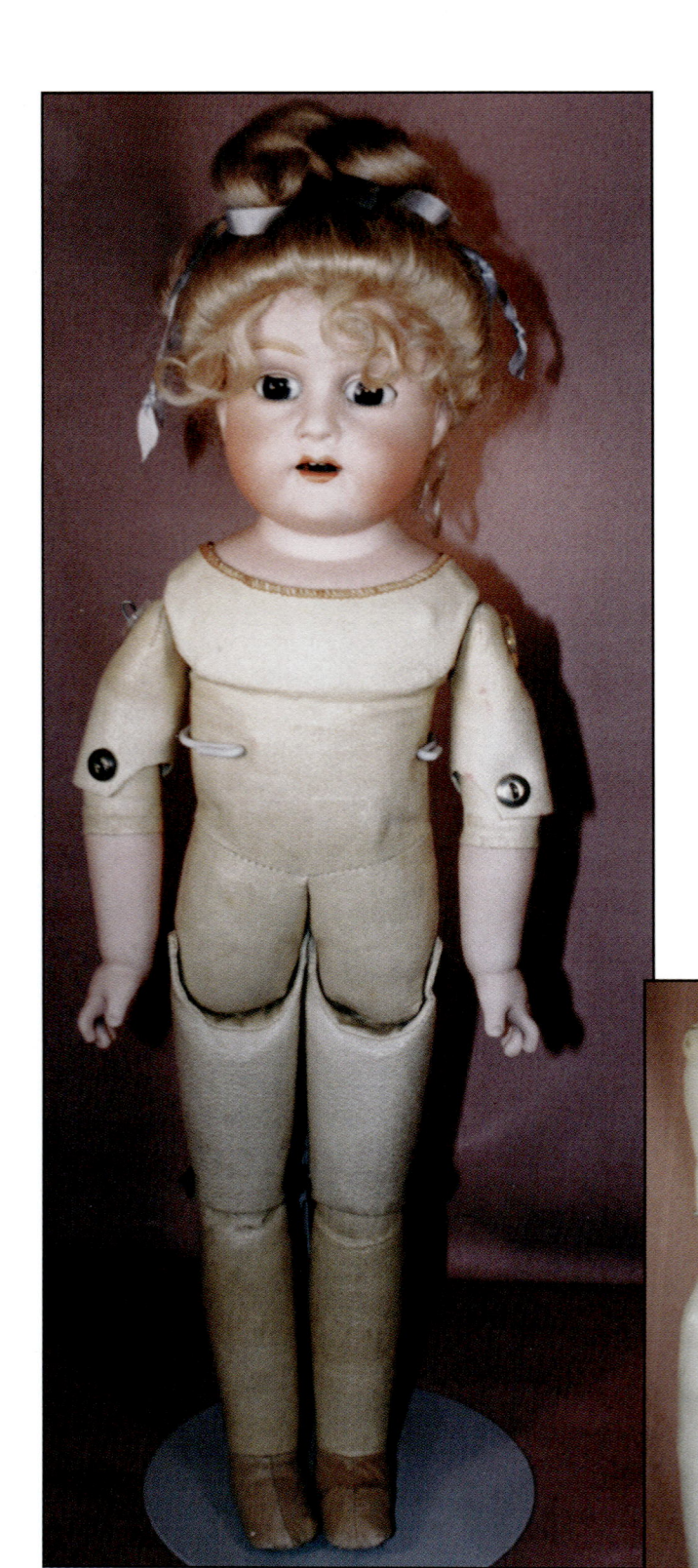

Repainted Bodies

Another concern for the doll buyer is whether or not an original ball-jointed body has been repainted. This also lowers value and desirability also, although not as much as a new body paired with an old head. Inspecting the inside of joints is the best way to tell for sure.

Of course, always ask before you handle ANY doll. Disaster could strike if the stringing is in poor condition. It is best, in buying situations, to ask the dealer to show you the doll. If you are in an auction preview situation and no one is on hand to show the dolls, try to examine the doll on a surface without having to actually pick it up. If you must handle it more than that, always pick up antique dolls with one hand holding the head. If the stringing gives way, better that body parts should drop to the floor than the porcelain head!

At times we see the oddest things. I wouldn't expect to find a good antique body paired with a reproduction head. I once made a lucky purchase of a "new" doll, a doll which had a new porcelain head that had been put on a valuable antique body. I was happy to pay the modest price to acquire the body and simply tossed the poorly done head into the donation box!

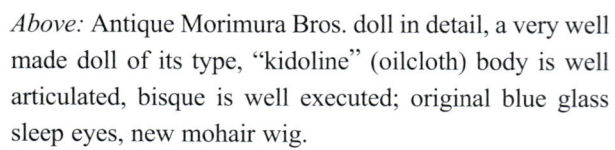

Above: Antique Morimura Bros. doll in detail, a very well made doll of its type, "kidoline" (oilcloth) body is well articulated, bisque is well executed; original blue glass sleep eyes, new mohair wig.
Middle: Preparation of a new vinyl body, the stuffed torso.
Right: New upper and lower arms are assembled and sewn to top closure.

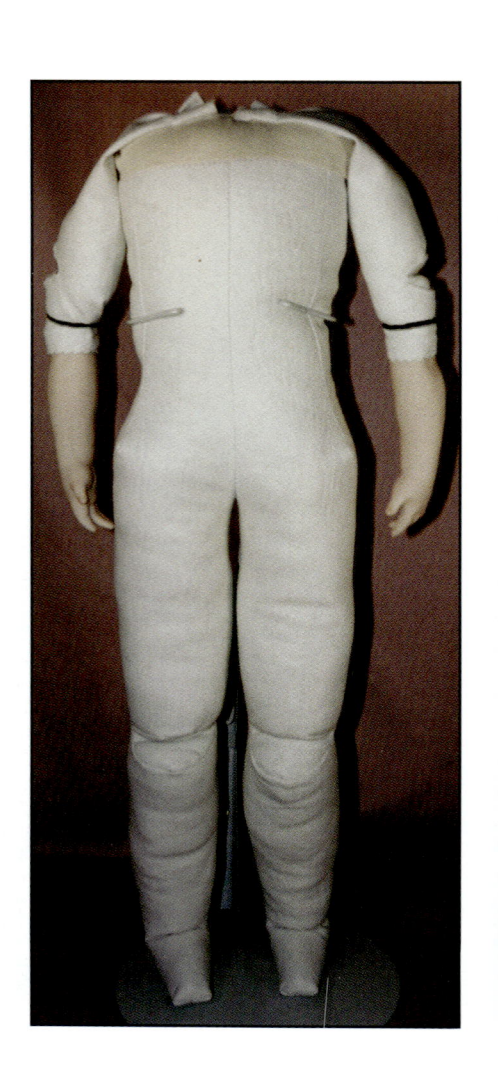

18

Kid-Bodied Dolls

Commonly found in doll collections and often a first antique doll purchase (possibly because of somewhat lower prices than the ball-jointed dolls) are the kid and kid-o-line bodied dolls.

Antique dolls which are on kid, kidoline (oilcloth), or cloth bodies require a different type of examination than the ball-jointed dolls do. It is probably not realistic to expect these kid/cloth bodies to be in their original state, although some really mint ones are found from time to time.

Kid Body Inspection

A cursory "feel" through the clothing will usually alert an experienced collector to body problems. Of course, leaking sawdust is also a good clue! Undressing these dolls is a must. Often, kid bodies which need repair will be taped or have strips of leather glued over tears and other holes. These dolls must be handled VERY gently...for the reason that while the leather may be in reasonable condition, the thread holding the seams together may be totally dry-rotted. Picking up one of these bodies may cause considerable harm. Again, it's best to let the owner of the doll do the handling.

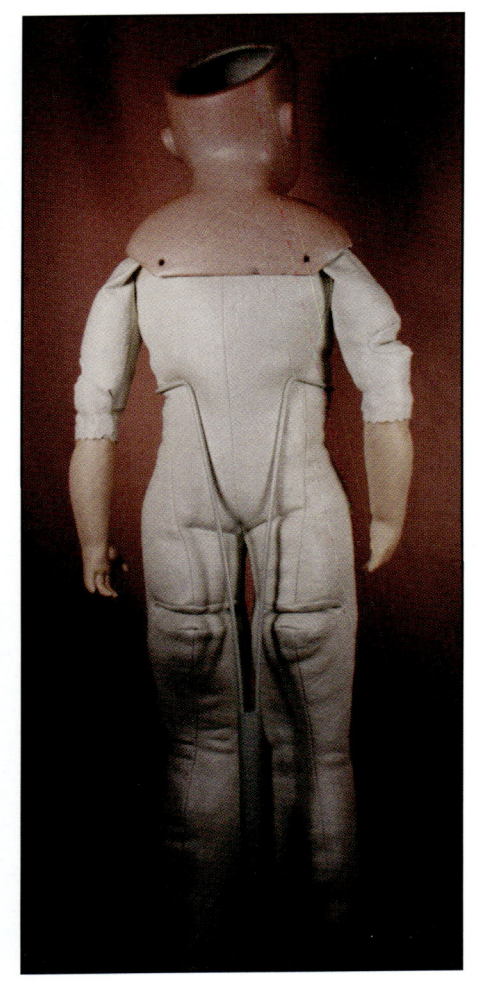

Back view of an antique "*Columbia*" head attached to the new body.

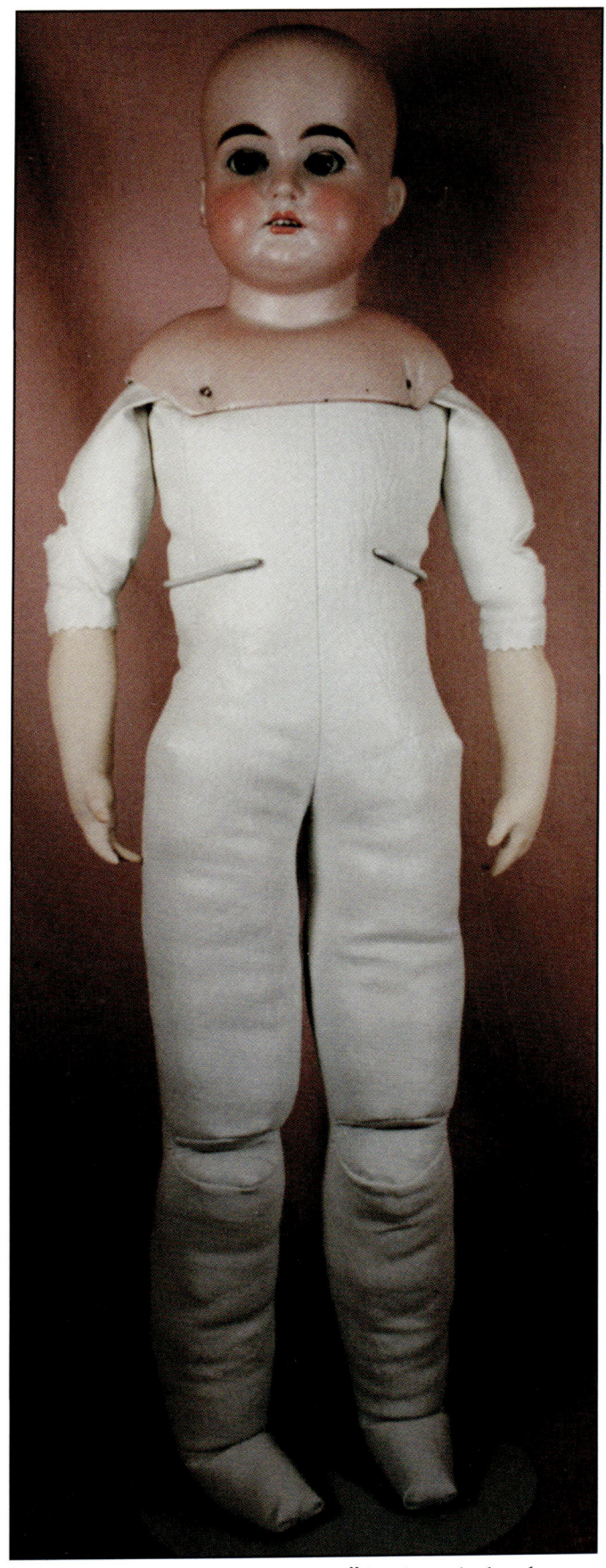

Front view of an antique "*Columbia*" head attached to the new body.

Comparing Kid and "Kid-o-line" Bodies

At first, the novice collector may have difficulty distinguishing between kid and kidoline bodies. Leather will have its usual grain. Close inspection of kidoline will reveal the cross grain of the fabric from which it is made; simply explained, a painted surface was applied to fabric to make this product. It was used for doll bodies as a replacement for the more costly leather, and to address the problem of sanitation. These bodies were touted to be washable, therefore more sanitary for children's play. Kidoline bodies usually survived better than leather or cloth, although repairs may also be indicated.

Price guides do not distinguish (at least at this time) between kid and kidoline bodies when pricing dolls. Since some mold numbers were offered first on kid bodies and later on kidoline, it would make sense to expect to pay more for the kid leather body version of the doll, all other things being equal.

Cloth Bodied Antique Dolls

The most common antique dolls using cloth bodies were undoubtedly the chinaheads and parians, although some of these are also found on kid bodies. Antique parians or chinaheads which have replacement cloth bodies usually are valued at about half of the price for the same doll with the original body. Of all the body types, the cloth bodies seem to be the most likely to have been replaced. If the replacement body seems older, say it was done about mid-way in the doll's life, and it is appropriate in scale and execution, value would seem to be reasonably intact. It is not reasonable to expect all cloth bodies from the 1870s and earlier to have survived.

Another defense of "homemade" cloth bodies on chinahead and parian dolls is that heads of these dolls were offered for sale. In the period that they were made, every household had ladies at the ready with needle and thread to meet practically all the family's clothing needs. Commercial clothing was not nearly as available as it is now, and certainly, not as economically. So for these moms, aunties, or grandmas, whipping up a doll body was of little challenge.

Some very oddly proportioned homemade bodies will be found, and are charming for their age and origin, if not accuracy. In these cases, the body, while not of "factory" origin, would certainly be the doll's original body.

More desirable than a replaced cloth body, however, would be the original body retained and covered in an "upholstery" style, with appropriate fabric. This is the preferred method to saving original cloth bodies in fragile condition. An example of this type of repair is shown in Chapter Four.

Wig Choices

The choice of wig should be appropriate, and is a consideration in determining the doll's price.

Technically, all antique reproduction dolls should have natural fiber (human or goat hair) wigs, since these are what the original dolls wore. As synthetic wigs are so much more affordable and available to the dollmaker, expect a lower price when these are used. The wig is one feature easily up-gradable by the new owner.

Antique dolls would preferably be found with the original wig, in the original set. This is becoming less and less likely as these dolls get older and older. Once in a while, a well-stored antique doll in amazingly pristine condition shows up, but it's a rare occurrence.

Original wigs in good condition, but soiled and in disarray may be carefully cleaned and reset with usually excellent results. When no wig is present or a hopelessly tattered wig remains, then a new wig in natural fiber is the appropriate restorative measure. With a replacement mohair or human hair wig in an appropriate style and pleasing color, antique dolls retain their usual value.

The "wig subject" is further, more thoroughly, explored in Chapter Six "Wig Care".

The Common Question: Does Restoration Lower Value?

Early on in my doll repair endeavors, a client came to me one day with a nice old antique bisque doll. She was quite soiled, with very loose stringing (The doll, not the client!!).

"The dealer who sold me this doll said it should not be cleaned or restrung", the lady related to me. "She said that doing this would lower its value. But she looks terrible! What should I do?"

So I took my time, phrasing tactfully the response that these "maintenance" kinds of "repair" certainly would not lower, but rather enhance, her lovely doll's value.

Then there was a gentlemen one day (we can't let the men be blameless in this!) who brought a doll to the shop which had no eyes. "Now, Miss", he said, "If you put eyes in this doll, will it lower the value?"

"Sir, what value does your doll have with no eyes?" I replied.

The point to be taken in any similar case is that value is already diminished if the doll has serious problems, and can only be enhanced with the proper corrections. Or to speak more bluntly, if your doll needs repairs, it already has some lessening of value, right?

Now, many years later and with less time left in life to spend worrying about subtle niceties, I reply to these kinds of questions, "Dirt is not a virtue, and neither is laziness!"

And now, dear readers, please take these words in the friendship and humor with which they are intended.

Careful Handling of Antique Dolls
Doll Etiquette

To avoid confusing the reader, it should be mentioned, that doll etiquette refers to the manners of humans, not their dolls! The dolls are always very well behaved.

Rather, we are concerned with, and will mention here, a few rules for those who will be handling and caring for dolls, particularly when the "human in question" does not "own" the doll.

And so, The Rules!

1. We do not really "own" our dolls, but are the temporary caretakers until they eventually come to be in the stewardship of the next human.

2. We owe it to the doll to "do no harm", i.e., unnecessary and improper irreversible repairs.

3. Never handle another person's doll without first receiving permission, whether as a visitor in a collector's home or at a dealer's table at a show.

Careful Handling of Antique Dolls

A prime consideration when handling old (or not-so-old) dolls is, simply, whether they can safely be picked up without falling apart. This is especially a concern with old dolls which are strung, although problems can arise with cloth or kid-bodied dolls as well. It is wise to have a firm policy about this, especially if you are looking at a doll at a show or might be repairing dolls for others. Assume the doll to be quite fragile; never pick a doll up by the arm or by the torso only.

In the case of antique bisque dolls with sleep eyes, it is necessary before handling the doll to make sure that the eyes are intact and not loose inside the head. If the wig is still glued in place, turn the head gently to see what position the eyes are in. If the eyes seem firmly in place, it is safe to handle the head gently. Never shake a doll to try to make the eyes work. If sleep eyes seem stuck, resist the urge to shake the head or to try to open them with any kind of a tool. Safe methods for loosening them will be presented.

When handling china-fired heads, whether old or new, it is well to remember that scratches and scuffs can occur, often caused by jewelry that we are wearing. Scuffs to "high" points of cheeks and nose tip can be caused by placing on an unprotected surface while doing eye work or stringing. It is best to remove all rings and wrist jewelry, especially any containing hard stones such as diamonds, and bangle bracelets. Always work on a padded surface, using old terry cloth towels, for example. In our household, everyone knows which towels are the "doll towels"; they even have their own hamper!

"Work" Storage of Antique Dolls

While the repair procedures are in progress, store your doll, or the parts, as common sense would indicate.

If a doll has loose parts or loose stringing, it's best to place it in a padded basket rather than in a stand. If you anticipate working on more than one doll at a time, it's wise to label (on little slips of paper, taped to basket) each one. It is easy to forget what goes where if your work is interrupted for a day or two!

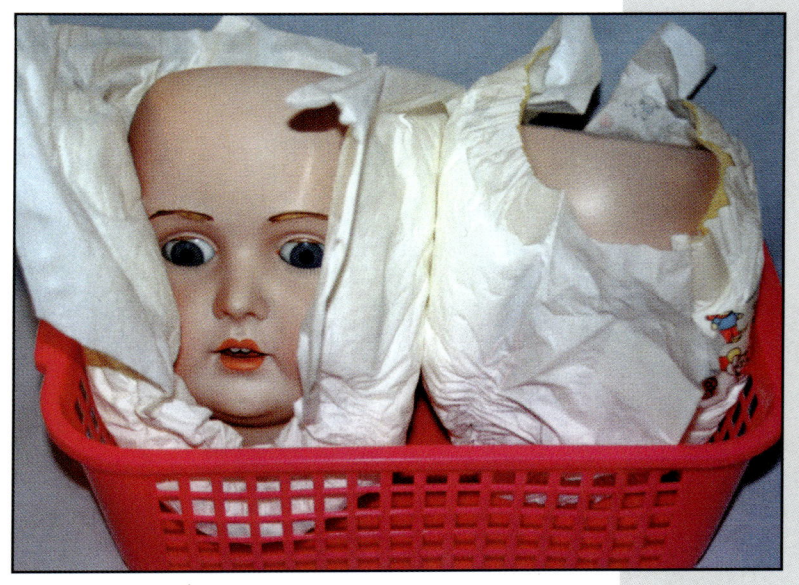

Antique bisque doll heads rest safely, wrapped in disposable diapers and tucked into a basket.

How to Choose and Use Doll Stands

At the risk of insulting experienced doll people, it seemed wise to include a couple of comments about doll stands (I've seen so many people struggle with these things!).

First, use stands with vinyl-coated stems so oxidation of bare metal stems in years to come will not stain the clothing or body of the doll. Check the weld of the vertical "stem" to the base to make sure it is securely attached. Remove the wire V-shaped stem from the stand base and pull each side to enlarge the V shape slightly. This will make it a little tighter in the base and it will hold your doll up better, especially if she is a little heavy.

Secondly, (always!) hold the antique doll by the head with one hand and place the V-stem at the doll's waist/chest with the other hand. Place the stem where

Chapter 3

it will "seat" the best; often this is not at the back of the doll, but at the front or side, depending on the shape of the doll's seat (see below.)

Check to make sure that the V-stem goes all the way through the lower "tab" of the stand stem.

If the stem doesn't reach down far enough, the doll may not be secure, and may tip about in the stand. If this is the case, try a larger size of stand. Doll stands are available for dolls from a mere 3 to 4 inches to 36 inches, so you should be able to get a proper size for most of your doll needs.

Displaying larger antique dolls, which are quite heavy, can be problematic, even with the wide range of doll stand sizes available.

Handling Dolls During Cleaning

Careful handling of dolls during cleaning is important to avoid breakage of these valuable old bisque heads.

Take hold of the head with one hand and support as much of the body as possible with the other hand. This sounds very elementary, but is necessary to avoid unfortunate accidents. Lay the doll out on a padded area which has been prepared using an old clean towel or other such material. This will not only help prevent breakage, but also little scratches or scuffs to the bisque parts.

Working on a doll that needs several types of repair, such as cleaning, eyework, hair work, and restringing, requires advance thought as to the type of workspace and work storage that are needed so the work will progress easily and safely.

Careful Handling While Dressing

Dressing a doll would not seem like an occasion when the doll might be damaged. Any manipulation of an old doll needs to be done very gently, and this of course also applies to putting on stockings and shoes as well as the clothing.

In particular, old kid and oilcloth bodies are susceptible to damage when handled. Fabric or seams can tear at the slightest touch, arms and legs can come off from their fragile attachments, and heads may be loose and drop off if the doll is picked up horizontally by the torso.

Therefore, so extreme care must be exercised. When putting stockings on a kid bodied doll, roll the stocking down to the ankle, and stretch it out. Then ease it carefully over the toes, and

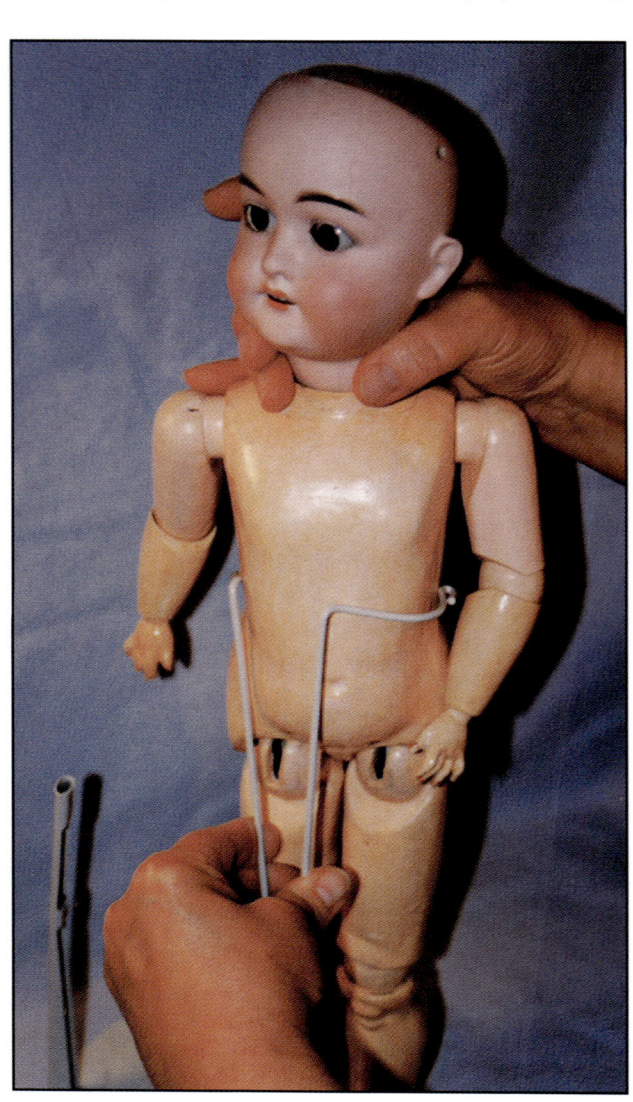

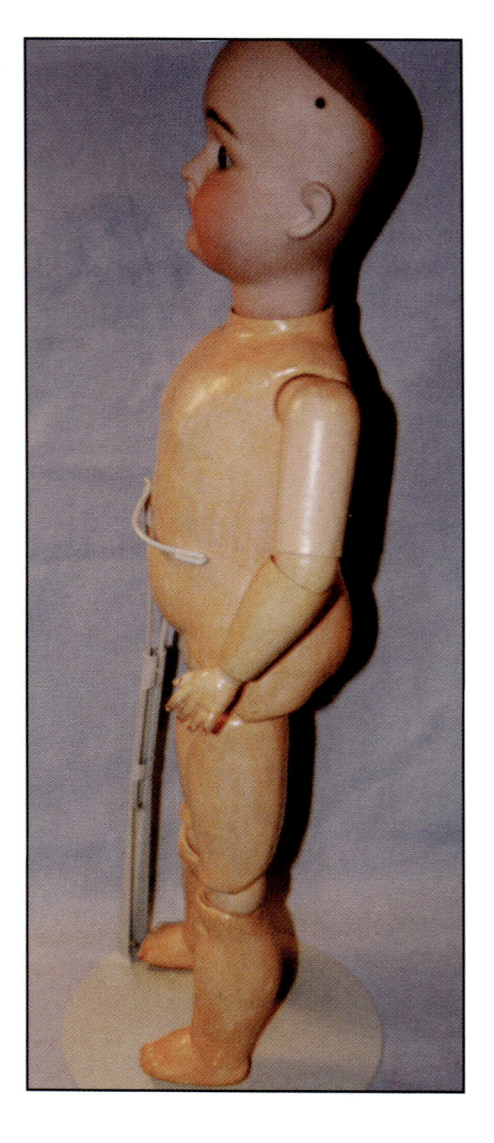

Make sure that the V-stem goes all the way through the lower "tab" of the stand stem.

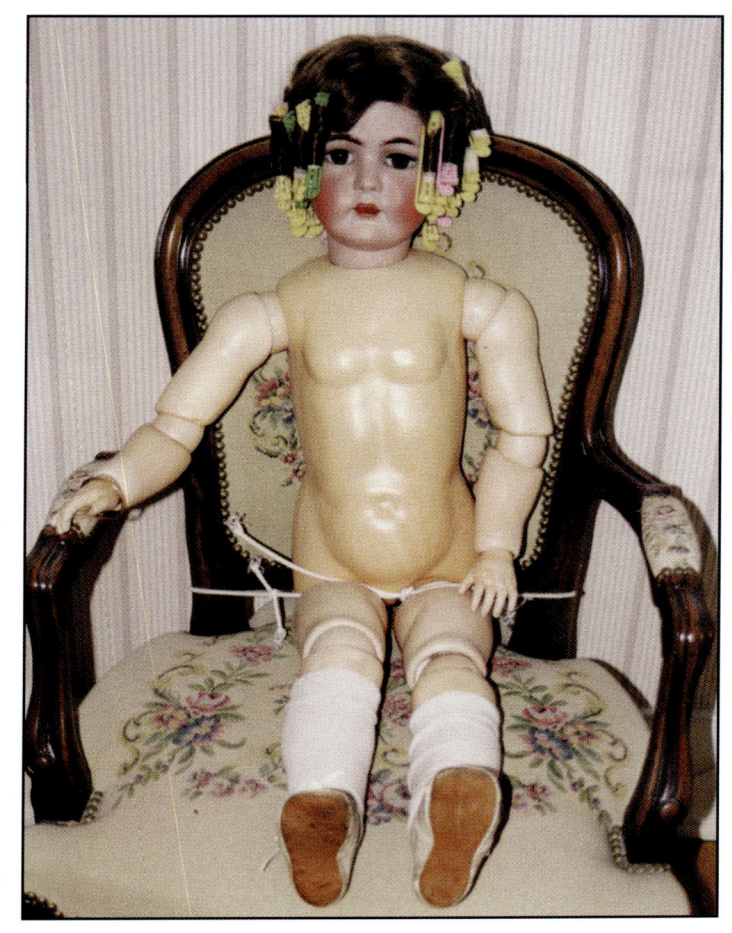

This 36" tall Kämmer and Reinhardt has been cleaned and re-strung, and her wig has been washed and set. She now waits quietly (does she have any choice?), tied up in a pretty chair, until her clothes are ready.

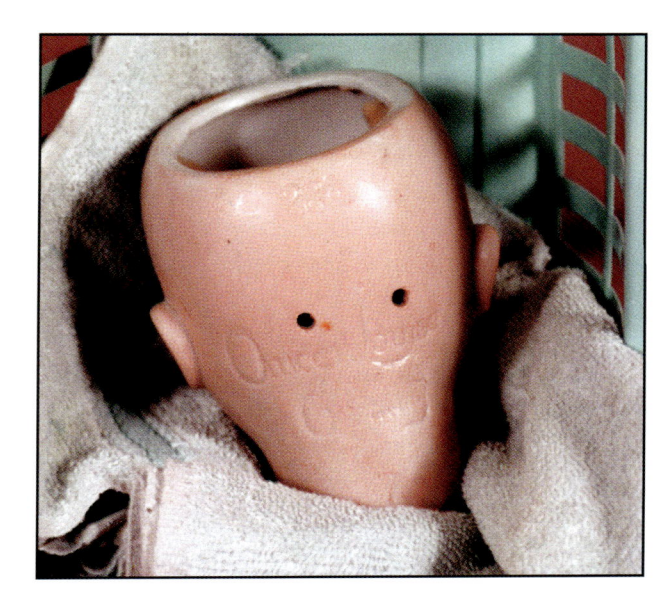

Antique *Queen Louise* head, back view.

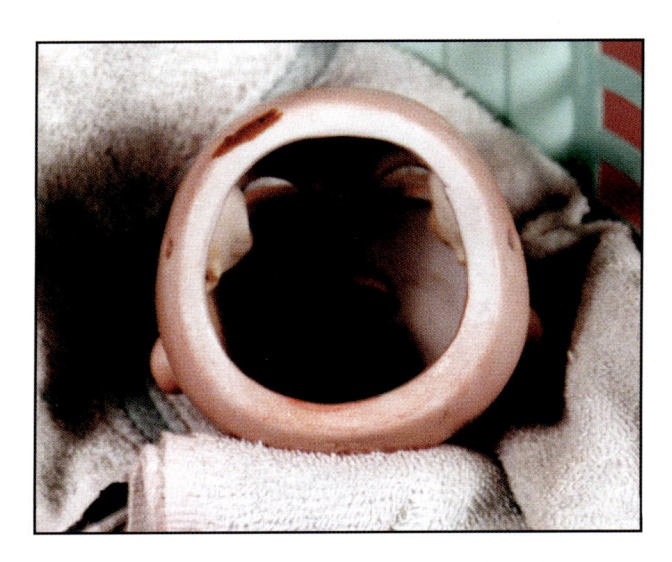

Interior view of *Queen Louise* head, showing plaster "eye pivots".

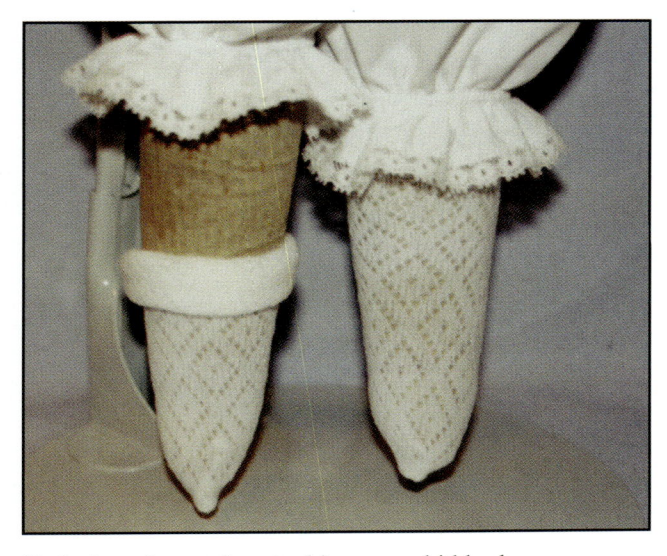

Technique for putting stockings on a kid body.

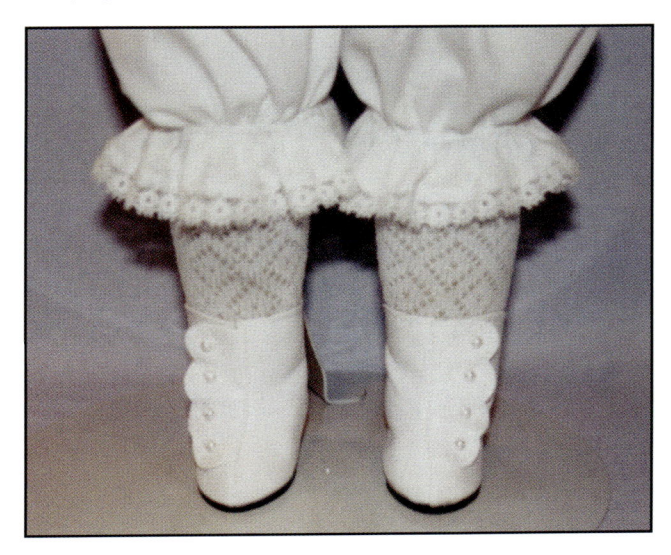

Dolly with completed footwear.

23

Organizing the Repair Work Schedule

The next consideration, if the doll is in need of some repair, is whether to repair first, or clean first and then repair. Usually it is advisable to do the cleaning first, unless the repairs needed cause the doll to be so fragile that even gentle handling will worsen its condition. This would especially be a factor when old kid or cloth bodies are decayed and/or leaking sawdust or other contents.

It is suggested that you read a little ahead in each section for these clues to timing. This may not seem important if you are working on only one doll, but if you get three or four going, as many of you will do, it can get more complicated.

For example, choose to do intricate work when you expect to have some un-interrupted time and no other pressing chores or errands. This way you can do all eyes one day and all stringings another. It does make good practical sense, because then you can get out all the appropriate tools and supplies once, instead of several times.

What to do First? A "Repairs" Overview

Imagine, as in the photo below, that you have such a "basket case" in front of you. A number of challenges are presented by this antique German bisque doll. Her eyes are out and broken, her wig is missing, body badly stained, she's all apart, and her clothes are, well, a disaster. She does have a cute little bisque "friend" in the box with her. So, where to start?

A doll repair agenda goes something like this:
1. Do the body.
2. Do the head.
3. Put them together.

A more detailed organization of tasks would be:
1. Clean parts first
2. Make necessary repairs
3. Assemble parts next
4. Costume next.
5. Glue wig on, add jewelry or accessories last.

Decide your own routine. Other than the natural progression of the work, there is no particular order intended in the sequence that follows.

A Common Sense Approach

Have one or two work baskets, size depending on the size of your doll. Mini-laundry baskets are suitable, and a plastic dishpan works well as no small parts will escape. One or two small gift boxes, the type that jewelry is sold in, are handy for storing the delicate glass eyes while they are out of the head.

Old terry towels or some other type of padding is essential.

Lay your doll out on the toweling, on a well-lit work surface.

If you cannot finish a task at one sitting, such as the cleaning, you can wrap the bisque head in one towel, body parts in another, eyes in the little closed box, and store everything in your work basket, safe until your next session. It's best not to leave the antique heads lying about unattended for any length of time, especially if there are children or pets in the household. If you will be restringing the doll, disassemble it before cleaning. It's much easier to clean joint areas, and of course, you will want to clean out the inside of the body. (We should someday hold a contest to determine the most unusual thing any of us finds inside dolls we are cleaning!)

Before Cleaning Your Doll: Some Gentle Reminders

It is crucial that, when attempting to clean a doll, one is knowledgeable about the material of which the doll is made. Irreparable damage can be done to the finish of the doll by using the wrong cleaning agent. A detergent which is safe and effective for cleaning fired bisque can damage composition or hard plastic dolls.

The best safeguard would be to try a small amount of the cleaner in an inconspicuous place, such as the back of the head under the wig or the bottom of a foot, if cleaning a body. With all of the experience and information one may have at hand, it cannot be known for sure what finishes were used, or what others may have done to the doll since its manufacture. Even with china-fired heads, there is some risk of removing color if the firing was not to a high enough "cone", that is , the temperature to be reached to complete the process. Kiln temperature may have varied, especially under earlier conditions, so caution is in order. Better to dwell on this point now than to have one doll suffer ruin.

Before we get into specific types of repair with specific methods (in Chapter Four), we will mention general types of repair problems. An important preparation to this work is to have the proper materials and equipment at hand. Many items you will already have.

List of Cleaners

Johnson's Jublilee Kitchen Wax, the cream, for composition bodies, leather, and kidoline.

409, a detergent, for cleaning bisque (test first to determine if the bisque was fired high enough); can be used full strength or diluted with water.

Isopropyl alcohol (rubbing alcohol), removes wax from glass eyes and paint from brushes (take care not to spill on painted parts, such as composition bodies!) Do not use on brushes during painting, but only to clean them when painting is finished.

Dishwasher detergent, crystals or gel, for really soiled fabrics.

Hydrogen peroxide removes (recent) blood spots from fabric, handy when you pierce a finger and bleed on your work! Also good for treating the injury to your finger.

Dawn, for washing synthetic wigs.

ERA, for dolly laundry.

Clorox-2, for stubborn dolly laundry.

Rubber gloves and old clothes for you, things can get messy.

Magnifier glasses.

Old towels, cotton swabs, cotton balls.

Materials NOT to use for Cleaning Dolls!

Petroleum jelly.

Vegetable shortening.

Cosmetics (too many unknown ingredients).

Hair spray, mousse, gels, etc (attract dirt, may attack synthetic materials).

Denture cleaning tablets (just an oxygen bleach, use laundry products).

Lighter fluid.

Shoe polish.

Cleansers which contain pumice.

Chapter 4

Doll Repair
The Cleaning of Antique Dolls

In discussing the cleaning of antique dolls, china as well as bisque headed dolls will be included, because the methods are similar. Other head materials found on antique dolls, such as papier mâché, wax over papier mâché, and wood will be discussed with the cleaning of composition, as similar methods are used.

To clean antique bisque or china heads, a solution of a mild detergent in water is safe and most effective. It is assumed that the heads were high-fired. Always test in an inconspicuous area first. Generally, the antique French and German heads will be fired adequately, although one can find exceptions. Some bisques made later, from the 1920s to 1940s, can pose cleaning problems. Often the paints were not fired on, and can be removed even by dry rubbing, so cleaning these would not be possible.

If you are working on a doll which will need to be restrung, it is best to disassemble it before you begin the cleaning.

Disassembling the doll

The head is best removed from the body by having a helper hold it while the stringing cord is cut. Remove the wooden neck "button" and, if in good condition, save to reuse. Unless you will be working on the head immediately, wrap it in a clean old towel, and position it with the eyes at rest, open. This is the famous "face down, eyes open" rule, which you will hear more than a couple of times! This is the safest way to store the head until you will be working on it.

The body can be completely disassembled later; it is important for the beginner to note any differences in right and left on the upper legs. The upper and lower arms are usually identical, and of course, hands and feet will scream out if strung in the wrong position.

Cleaning Broken bisque or china heads

It is also very important, before getting into work on the doll, to note any hairline cracks or other breaks or damage to the bisque or china. Until you establish, with the use of a magnifier, that there are no cracks or hairlines in a head, assume that there could be breakage. This policy will keep you out of all kinds of trouble! If cracks are found, and especially if pieces have been glued, these issues will need to be addressed. Glued areas may loosen during cleaning and present problems when stringing the head.

The approach then would be to soak loose any previously glued repairs, if possible, and re-glue to assure that they will hold up to stringing pressures. The glue line inside of the head can be reinforced by gluing down a thin strip of muslin, cut on the bias so it curves easily. The fabric will absorb the glue better if wet first with water. A good quality water soluble-white glue is the best glue choice. It helps to use a "tacky" variety, as the pieces will set up quickly. This type of "repair" works well on clean breaks and is reversible, should another type of bisque repair be sought in the future.

Cleaning the doll head

Before attempting to clean the doll head, the type, the condition and operation (movement, if any) of the doll eyes must be determined.

If the eyes are absent and will need to be replaced, any plaster which was holding them and needs to be soaked loose and removed during the cleaning.

This antique doll head, quite soiled, is fully immersed to begin loosening soil and glue since the eyes are peeling paint and wax and do not fit well, they are part of the soak-it-loose process.

If the eyes are present but out of the head, it may help in if the plaster "curves" which hold them in place are saved. Be sure to mark these pieces for "right" and "left" sides of the doll head.

If the eyes are intact and working, or set fixed,

care must be taken not to get the plaster holding them in place too wet during the cleaning. Soaking the plaster is how it is removed!

If eye repair is needed, the eye repair section should be studied before doing any cleaning. The eye repair can be started, as some of the stages may require drying time, and other repair work can then be planned.

If the eyes are out of place, usually one of the plaster curves is in place and one is loose, or absent. Save the loose piece, if found, for use in resetting the eyes. It should be kept dry, so it should not be cleaned.

It's safer to clean bisque heads in a lined container, rather than over an open sink.

It is important to mention that, when wet, antique bisque heads become quite slippery. Again, laying the head on a pad in a small basin is best. Spritz on a solution of detergent and water outside the head. Take care not to get the eyelashes wet, as they may come loose. Clean with a soft cloth or cottonballs, and yes, cotton swabs for ears and nose. After all, this is someone's baby!

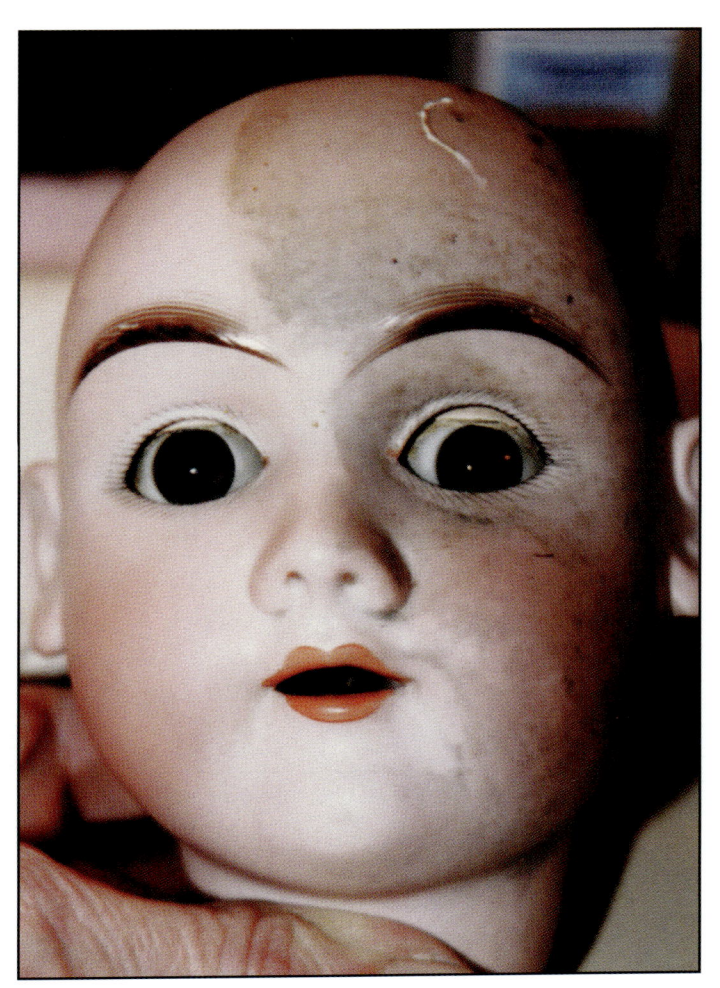

Since the eye pivots may be loosened by too much water, this head is cleaned by sponging and wiping rather than by soaking. Cleaning only half of this antique bisque head with 409 and a cottonball shows immediate and remarkable results. What a beautiful face is under all this dirt! This doll, a charming antique Max Handwerck, will be shown in Chapter Five, under "Adding Replacement Eyelashes".

The inside of the head is also carefully and thoroughly cleaned.

This very dirty head needs, and can have, a thorough soaking, because the eyes are out and the eye pivots also need to be soaked loose.

If there is glue residue from the wig, it may not easily dissolve. The glue area can be wrapped in a wet paper towel, a narrow band of it around the crown of the head. If no eyes are present, the head could be placed inverted in a small bowl of warm water. It may take several hours to soak the glue loose. It is advisable to wait for the soaking to take effect rather than scrape or scrub the head, possibly scuffing the surface.

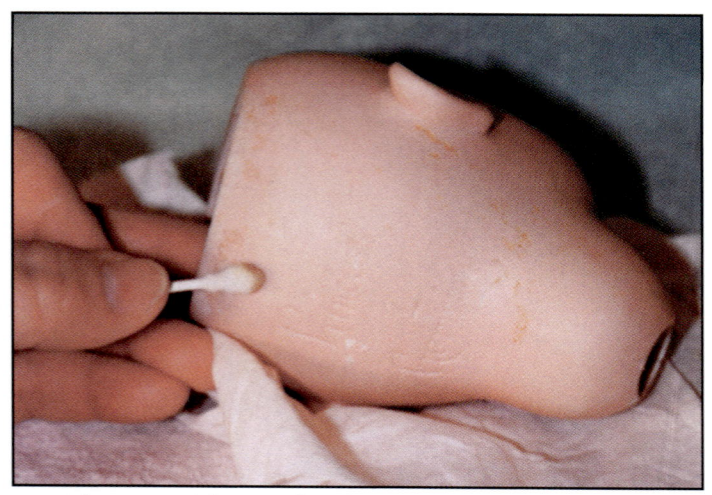

After several hours of soaking, this head still shows some glue residue (darker yellow color), which is coming off with gentle pressure from a moistened cotton swab.

Clean inside the head very carefully, using a cottonball wound around a chopstick or piece of dowel rod. A sable paintbrush works well around and behind the eyes, and limits the amount of moisture. When the cleaning is satisfactory, rinse the head well, and pat dry. It still needs to air dry for a couple of hours, as unglazed porcelain absorbs water. If no eye work is needed, the head can be set aside to dry while the body is cleaned and re-strung.

The (Sometimes) Dirty Inside Story

Getting into dolls (inside of them) can be very interesting. Strange things are put into dolls, we'd like to think by children, but that's not always the case. You will probably find a few cracker crumbs or some Cheerios someone fed them for lunch, but we have a few items that were a little less expected.

I once disassembled an early (probably 1910 or so) all composition which was a rather large and masculine looking boy. The hooks which would have been on the arms for stringing were broken away and someone with perhaps more determination than resources had used a large corn cob to sort of "screw" the arms into place. As I knew the owner would have a good chuckle over this, I saved the corn cob to give to her when she called for the doll. She had a good laugh, and said, "That was Grand-dad. He had this doll stored out in the barn for years."

Another composition mama doll, cloth bodied, had a tinker toy where her heart had been. I wonder if her young owner is now an accomplished cardiologist!

One doll came with her own allowance. An elderly lady brought in a medium-sized bisque-headed doll, wanting to know if we could repair the eyes. To assess the eye situation, I carefully removed the wig and pate, and from the head fell a folded-up ten dollar bill. The woman shook her head and said, "I just don't know about my friend (the doll's owner). She is quite up in years and I have been helping her clear out her house. I am finding money everywhere!" I said, "Well, at least her doll has a little spending money!"

More "Inside" Stories

A young mother with two rather active young sons came in one day with their Teddy Ruxpin. This is a large bear with electronics which operate a cassette tape, and as the bear "talks", his mouth and eyes move. Teddy works all right, she said, but there is a new funny noise when the tape is operated. Checking into Teddy's mouth, we found that the boys had fed him quite an adequate lunch, a number of large nails! When the lady returned for the bear, we mentioned the nails, thinking she would want to limit Teddy's diet in the future. "Nails?" she asked. "That's quite impossible! Surely you don't think my boys could have done that!"

The "topper" in doll cleaning stories has to be the find in an all composition girl doll which was about 24 inches tall. We took her apart to clean and re-string her, and found we had to remove quite a quantity of insulation from inside her torso. Why would anyone put this inside a doll, we wondered. In the next handful was our answer, a quite dried up and very dead...mouse! About five minutes later we all let out a scream! Delayed reaction, yes, we had to think about it first. A doll torso would make cozy quarters with the right "bedding". This doll's owner was a young local medical doctor just setting up a new office. We decided that the better part of diplomacy was...silence..and this remains our awful little secret to this day.

Can you top these "inside" finds? We'd love to hear if you do

Doll Teeth

If the doll is an open mouth, it probably had teeth. Dolly face dolls usually had four, sometimes six, small upper teeth, backed up visually by usually nothing more than a small piece of red paper. In the baby dolls, usually only two upper teeth were used, often with some type of molded tongue which was set in behind. Many people do not notice that their doll is missing teeth until it is mentioned.

Original teeth were done in basically two ways. Commonly they were thin white plastic shapes, squared at the top and rounded at the bottom, glued in place. Often one or two of these will be missing. It is really difficult to craft matching teeth from virtually any material, and it is not recommended.

Besides teeth which were put into the doll head, doll teeth were also done as part of the head mold. If there are chips to this type of teeth, there is no way to remove and replace them. In this case, a professional porcelain repair would be the best solution.

To replace teeth of the installed variety, it is simpler to soak the remaining teeth loose and replace all with new plastic or porcelain teeth.

Plastic teeth are readily available, in sets of two, four, or six, and in three or four sizes. So it's very easy to arrive at the proper scale and number of teeth, as the material from which they are made may be cut with a small scissors.

Porcelain teeth are made from molds which are readily available from doll mold companies. They are of course small molds, with a few size and style choices. Porcelain doll teeth are easy to make, but of course require a kiln for firing.

If the eyes are out of the head, install the teeth first. It is easier to gain access without having to maneuver around the eye rocker. If eye work is not needed, the installation of the teeth will be a little more difficult in terms of access. On larger heads, it may be possible to work through the neck opening.

To install teeth of either type, use the following procedure:

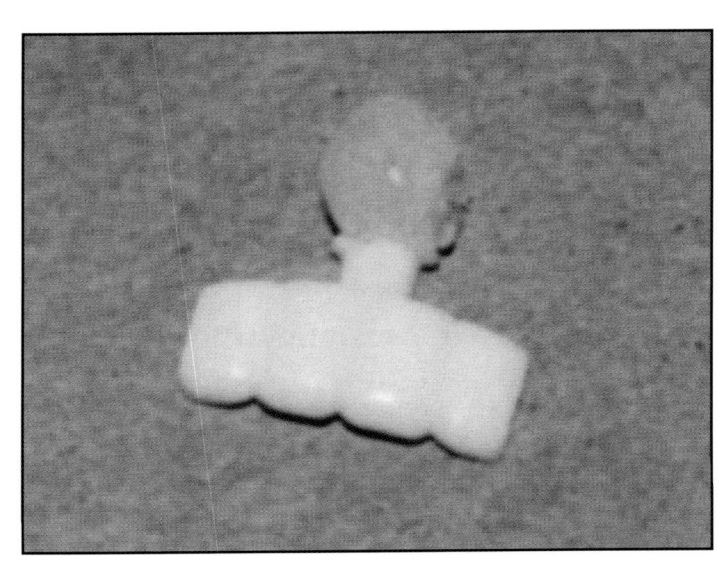

Roll a small amount of sticky wax into a ball, and press to the tab above the teeth. The replacement teeth are then pressed into place in the head, held by the sticky wax. Move them about to make sure they are centered, and not too low. When you are pleased with the look, cover the tab with a small amount of plaster. Position the head in a slanted position so that the wet plaster does not run down out of the open mouth! Allow a few hours of drying time.

When the plaster is dry, you can handle the head again. To decide if you wish to install a tongue, place your hand over the open crown of the doll's head. This cuts off the light and limits almost totally any view into the doll's "mouth". You may decide a tongue is not necessary. If you wish, a dark red paper tongue may be glued behind the teeth when they are dry. The use of felt for tongues in antique dolls should be avoided. As it was commonly used for this purpose in the mama dolls of the 1930s and 40s, its use will imply an incorrect assumption of the doll's age and type.

Ball-Jointed Body Repair

Ball-Jointed Body Terminology

It would seem logical that we standardize our terminology for doll parts, so that the directions are more easily understood by you, the patient reader!

Below is a sketch of the basic German ball jointed composition doll body. This is the type of body that you will most often encounter on strung antique dolls. Names for the various parts which we have developed over the years and gleaned from others are listed.

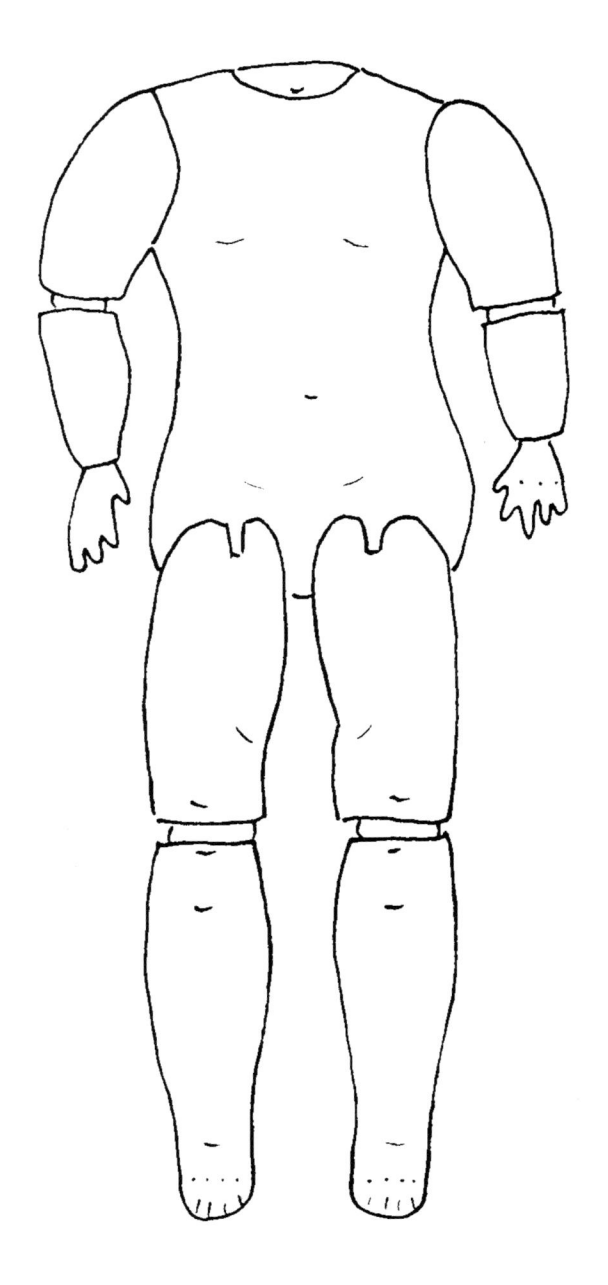

Cleaning the Ball-Jointed Doll Body

Doll bodies of the strung varieties are more easily cleaned when apart; if the body is to be re-strung, disassemble it, taking note of any peculiarities about right and left-sided parts. As noted before, hands and feet will be obviously right or left-sided. Upper and lower arms are usually identical. Upper legs will usually have a right and a left, usually curved out on the outside and indented or "dimpled" in the inner thigh. Ball joints for elbows and knees are usually easy to distinguish one from the other; occasionally, though, they are quite similar in size and shape. Taking note of the differences (or lack of differences) in the ball joints ahead of time will prevent a little headache when you begin to re-string the body.

Cleaning of composition and ball-jointed bodies simply requires a soft cleaning cloth or paper toweling, a small soft brush and a good cleaning agent.

A time-tested favorite cleaning product of many doll restorers is Johnson's Jubilee Kitchen Wax. This time-honored product was widely available for scores of years as a white cream which housewives used to polish kitchen appliances. It is now sold as an aerosol spray in stores, but the original cream, which is much preferred for cleaning dolls, is still available directly from Johnson Wax by mail order. Unique for its qualities of conditioning paint and removing soil, Johnson's Jubilee also leaves a nice sheen and is not unkind to hands. Apply the cream with a soft cloth or your hands; use a small paint brush or cotton swab to get into crevices. Even if the doll body looks clean, you will be amazed at the soil that comes off on your cleaning cloth. Use enough cleaner to keep the working area moist, and buff away while wet. A second application may be necessary if the doll is quite soiled.

Other products are touted for cleaning dolls and may also be tried. As with cleaning any surface, test a small area first. Be cautioned, however, that products advertised for vinyl dolls are not suitable for composition and ball-jointed bodies. The various materials (cardboard, wood, and mâché) which make up a ball-jointed body could be harmed by these products, which contain largely water.

Wherever doll people gather, cleaning suggestions will be heard. Some are not to be believed. Methods which should NOT be used include: applying make-up base to composition, using petroleum jelly, various cold creams, vegetable shortening, wood cleaners which are mixed in water, hair spray, lighter fluid...the list goes on.

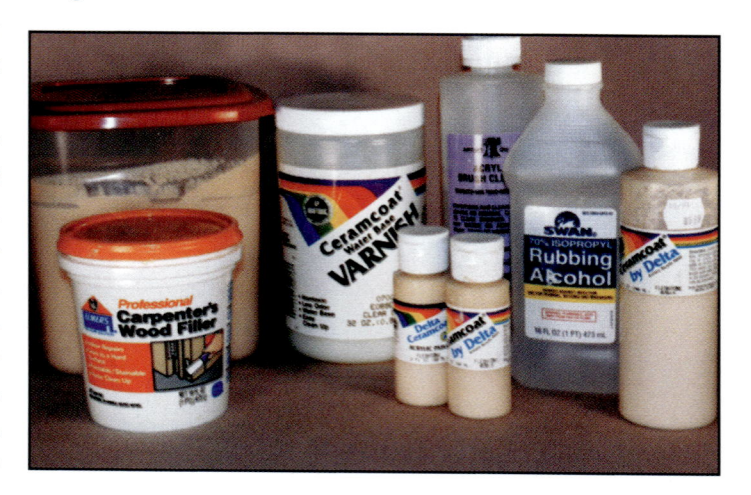

Products Needed for Composition Body Repair

Some good basic products needed for composition and mâché body repairs are shown below, and include:

Isopropyl alcohol (rubbing alcohol), for cleaning glass eyes of wax prior to painting

Carpenter's Wood Filler, for building fingers and filling cracks, it dries hard and drills and sands well.

Durham's Rock Hard Water Putty, for final fill-in of putty finish.

Ceramcoat Acrylic Paint..

Ceramcoat Acrylic Varnish, mix with paint to get a hard but pliable, very durable finish.

Acrylic brush cleaner, not so great for cleaning brushes, but recommended for thinning acrylic paints rather than using water.

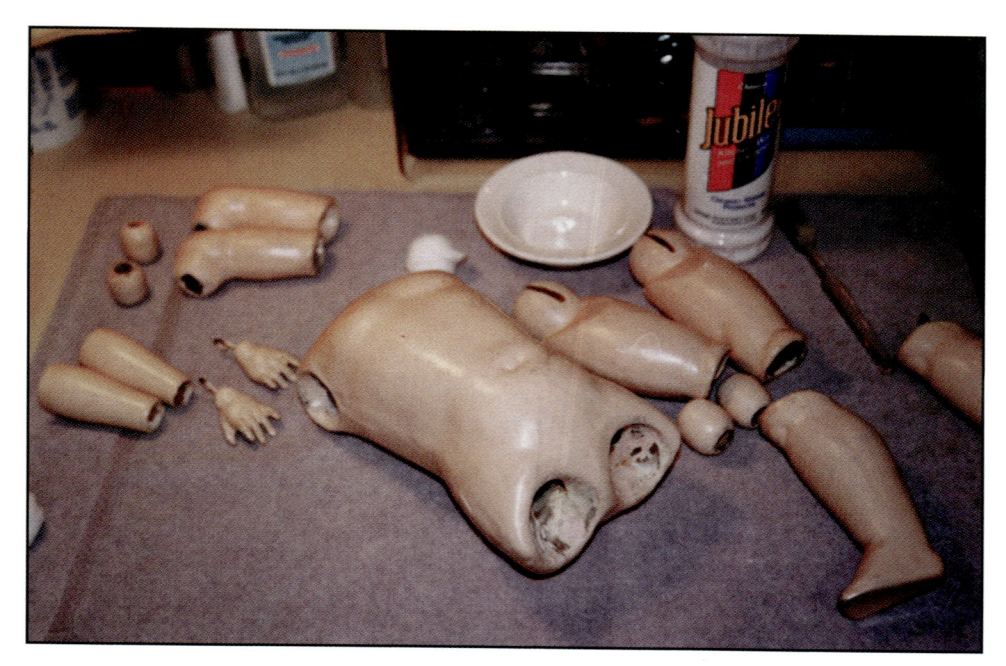

Preparing to clean composition/ball-jointed body parts.

Tools and other Supplies for Composition Body Repair

Spatula for the "putty".
Good quality paintbrushes.
Sandpaper.
Small dishes for mixing paints.
Bottles or containers with lids for storing mixed paints.
Two pairs needle-nose pliers.
Wire cutters.
Coat hangers may be utilized for wire to hang parts that are wet.
Gauge 22 steel wire for making doll fingers and other small shapes.
Wax paper, to cover tabletop and to lay small painted items out to dry.
Old terry towels, paper towels.

A Few Painting Rules

1. Use acrylic paints. There is almost no risk of acrylics reacting with existing paint surfaces and a wide variety of colors is available. They also store well for long periods of time.
2. Mix with a popsicle stick/tongue depressor.
3. During the painting procedure, brushes should be soaked in water or acrylic thinner. Use isopropyl alcohol to clean the brushes when painting is completed. If you use the alcohol between paintings, some of it may remain in the brush and affect the painted surface. Used for the final cleaning of brushes, it has ample time to evaporate before the brushes are used again.
4. Paints to which you have added varnish (such as for the eye dipping) should be refrigerated (in a sealed container) if you will not use it up during the day. Besides thickening, mold can grow in this mixture if it is kept at warm (room) temperatures for a long period of time.
5. If your acrylic paints need to be thinned (they do thicken over time), use acrylic thinner. Take note that oil paint thinner can not be used with acrylic paints.

Painting Repaired Areas on Composition Parts

To paint a small repair and match the surrounding old paint finish, try the following procedure:

1. After a thorough sanding, paint the repaired surface with acrylic paint mixed to match the original finish as closely as possible and allow it to dry.
2. Apply a "wash" which is paint-thinned to almost watery consistency with acrylic thinner. For this wash coat you use the same paint as in step 1, with perhaps an adjustment in tint to match the surrounding finish, especially if the old finish was shellacked and is somewhat yellowed. After applying the wash, allow drying time.
3. If you are doing fingers or toes, apply any "lining" that is present now. Then apply a coat of acrylic varnish and allow drying time.

Repainting Antique Bodies?

Many discussions have been held by doll lovers regarding the repainting of antique doll bodies. In general, repainting an entire doll body lowers a doll's desirability and therefore its value. I am often surprised how many doll owners want this done, as they feel that they want the doll to "look better". This is certainly understandable, but a scuffed original finish is more desirable than a newly painted finish on an antique body. Often, my comment to these doll people, who do love their dolls, of course, is to enjoy the doll as it is, once cleaned, wigged, and costumed. Minor touch ups of fingertips and toes, if done well, will go unnoticed, but even this little amount of repainting should be discouraged.

Repairs before Re-stringing

This is also the time to note any repairs which are needed so that the re-stringing will be sturdy. Check all hooks, making certain they are solid. Fill in and around any loose hooks with a good quality putty of a type that would be used on wood. Several newer products are water-cleanable and easy to use. A simple technique to use is, after applying about the right amount of putty, to smooth it with a wet (dip in water) paintbrush, or on larger areas, a finger works well. When dry, little sanding will be required. If desired, color touch-up can be done with acrylic paints of the type used in craft projects. Several good brands are available and come in a wide variety of flesh shades. Keep yellow, red, and brown paint on hand also, for tinting.

Missing Body Parts

When antique dolls are missing a part; an arm, leg, or hand, the best "cure" is to locate a nearly identical replacement part. Of course, finding an antique part which is the same in all dimensions and color is nearly impossible. It's not that doll hospitals don't have stocks of this type of item, many actually have a veritable horde of parts. The problem which arises is that you can look at a thousand antique doll arms and not one will be "close enough" to what you need.

A visit to the doll "bone yard" helps one realize just how many different shapes and shades of flesh there are. One wonders where all these pieces have been over the years?

Missing Body Sockets

Once in a while a doll body will be missing sockets on the body, either at the neck or arm openings. These were made of shaped cardboard and glued into place. It's relatively easy to replicate the basic idea.

Use unfinished cardboard of the type that you would find on the back of a tablet of paper. It needs to accept water and yet hold together when wet. Establish the general shape you will need by using the piece which will be fitting into the socket, either the doll's neck, shoulder joint of the arm, or hip joint of the leg.

To make a replacement for a missing socket, begin by making an approximate measure around the shape which the socket will receive, such as the arm (See below, upper left.)

Next, cut out the basic round shape, cut or drill a hole in the center (for stringing [see below, center] later) and then make kerf cuts into the (outside) circumference. Then soak in warm water for a few minutes. Since water would probably damage a composition doll part, wrap it in clear plastic wrap. The cardboard, while wet, can be shaped about the arm where it will fit into the socket and held by rubber bands until dry.

It is then placed into the socket opening (see below, right and a line is marked about the edge where it meets the body opening. Trim along this line, as needed. Apply a "rim" of wood putty just inside the socket opening.

Place the socket shape into the opening (see below, lower left). Using a long rod such as a chopstick, go into the body at an opposite opening and tamp the putty firmly into place about the socket. You will need to hold the socket in place with your other hand.

When dry (allow about a day), fill around any open seams with plastic wood fill putty, allow to dry, and sand as needed.

If the neck socket is missing, it may seem more of a task, but is actually easier. The opening is a little larger and easier to access. The neck of the doll head may be used to shape the wet cardboard (see drawing on page 33) without worring about moisture damaging the part.

When the socket repair is dry and sanded smooth, touch up with acrylic paint, matching the body paint color as closely as possible. Some of the old paint colors are very difficult to match, especially if a finish coat of shellac was applied. The shellac coat is usually an transparent amber or orange color, rubbed off in some places, so a uniform appearance is almost impossible to achieve. The appearance of the old shellac finish can be mimicked; suggestions for this procedure are found in the section about painting composition repair areas on page 38.

Strive for neatness and do the best you can in blending in the repair area. I try to do the least amount of repainting possible on old dolls. I console myself that someone with more skill than I in matching paints may always improve upon the job, with no ill results to the doll.

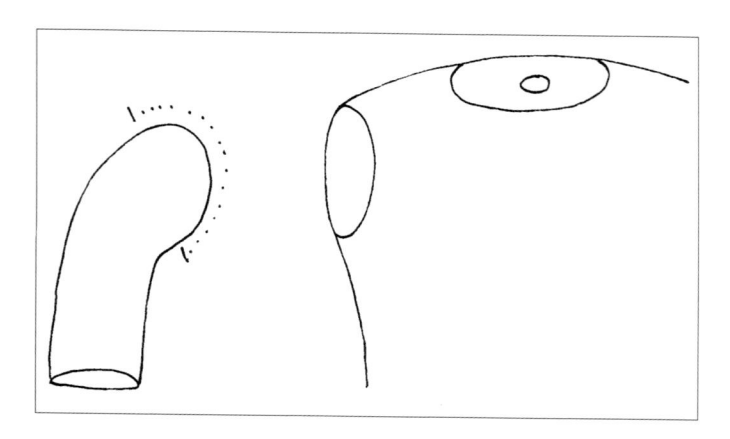

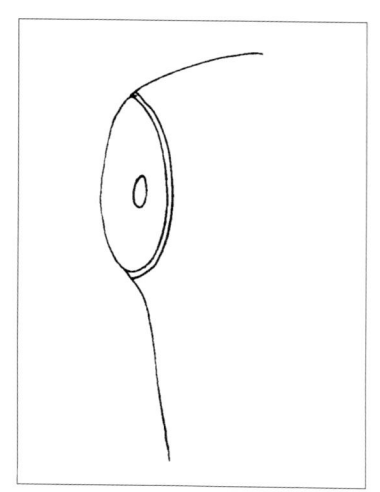

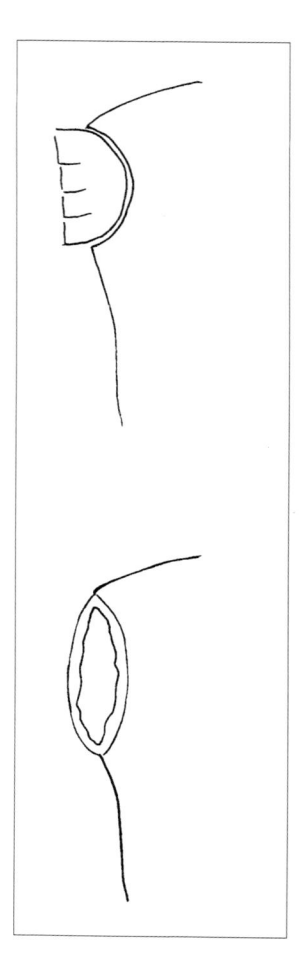

Body Sockets Present But Loose/Misaligned

A common repair needed in ball-jointed dolls is the re-setting of parts of the body which have come loose, or are fixed, but in a misaligned state.

Our first example is the ball-joint area of a upper arm.

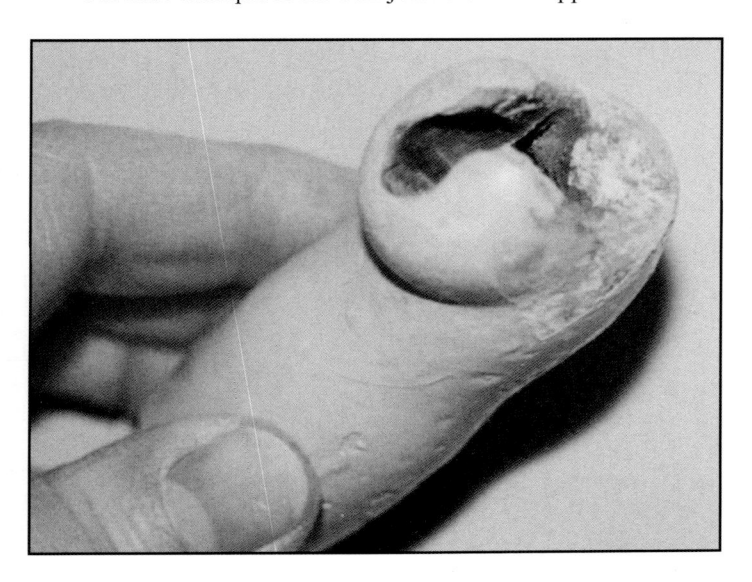

As seen above, part of the wooden ball-joint has broken away and is missing. The repair is needed for cosmetic reasons, yes, but also because if the body is strung as it is, the stringing may be caught in this open area when the arm is rotated.

To begin, a small piece of stiff brown paper is rolled into a tight tube shape to fill the opening and give a shape against which the filler can be applied (see top right photo).

The wood putty is then applied and allowed to dry (see right center photo).

Then the rolled paper is removed, twisting to release it if necessary.

The filled area is thoroughly sanded (see right).

Touch up the paint by applying a coat of acrylic paint, mixed to match as well as possible. When this coat dries, apply an acrylic wash to mimic more closely the appearance of the old finish.

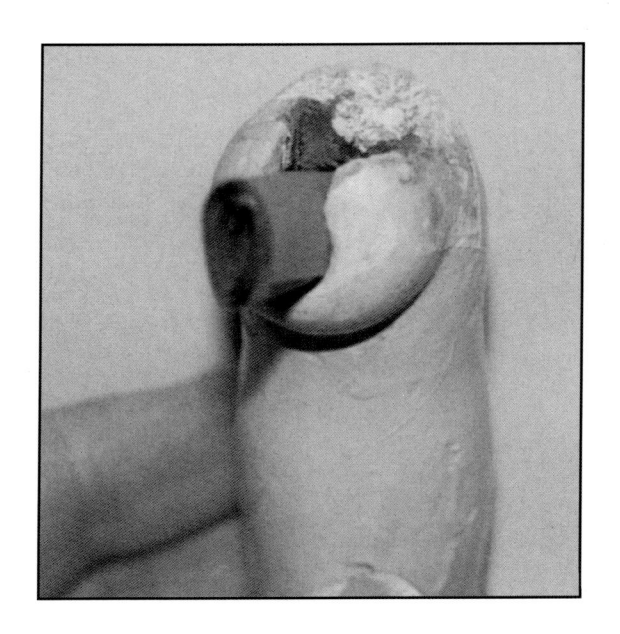

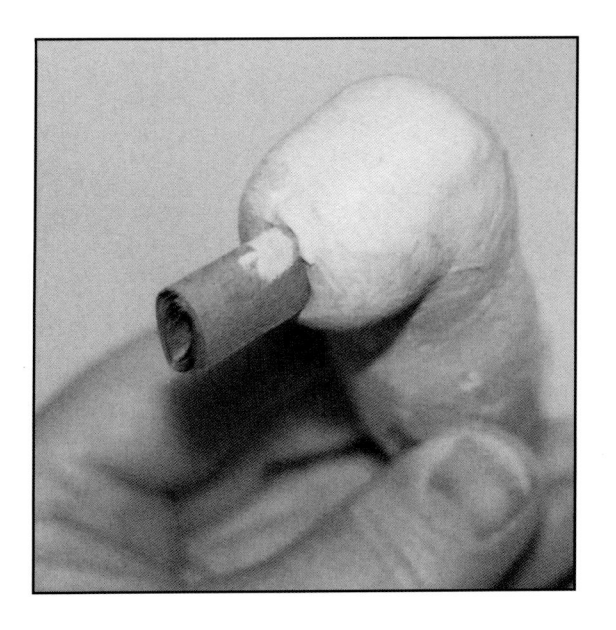

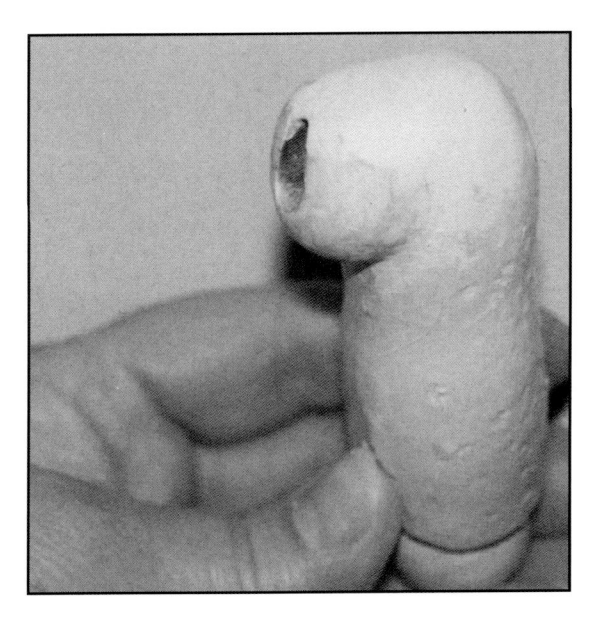

When this coat is dry, apply a coat of acrylic varnish, allow drying time, and the part can be strung. (see below).

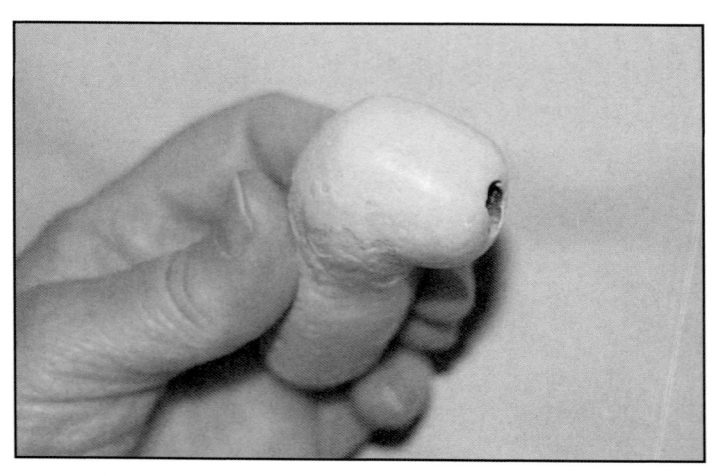

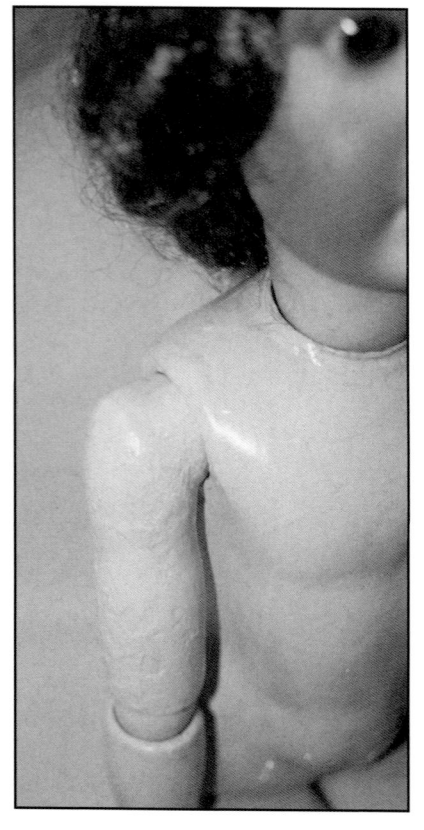

Example two is an upper ball-jointed leg, which has two problems to address. The wood ball-joint is out of the cardboard leg (see photo below). In addition, the cardboard portion is worn down, and the top edge rolled over, preventing the ball-joint from seating properly. As a result, with the ball-joint placed into this position, the leg is noticeably longer than it should be, compared to its mate (see above right photo). The leg needing repair is to the left in the photograph.

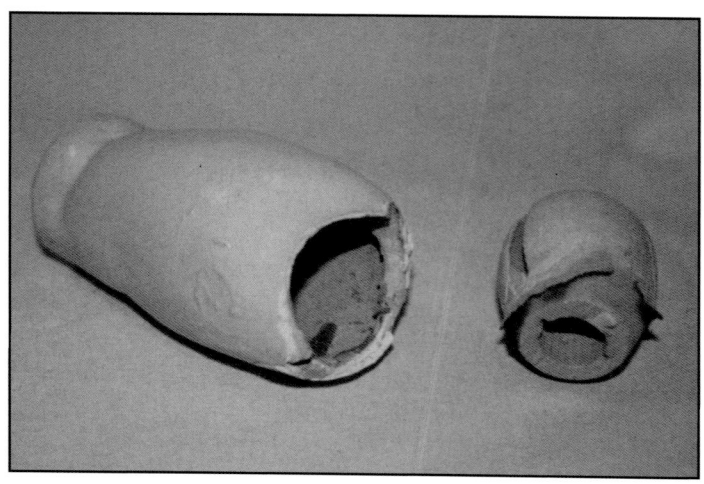

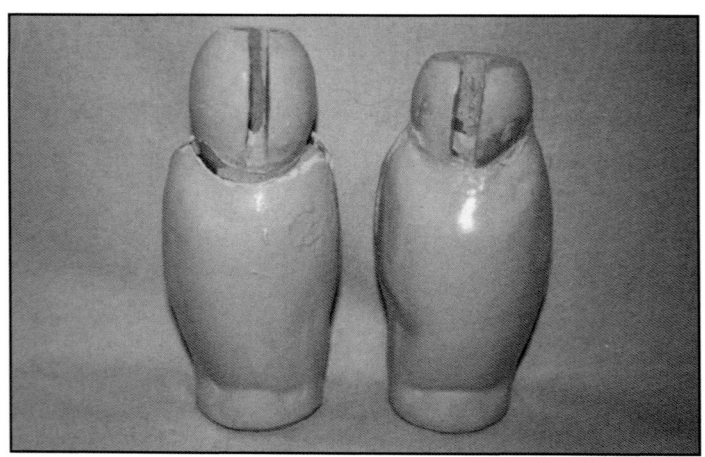

The first step is to correct the length problem. The interior of the cardboard leg is sprayed lightly with water. When it can be unrolled, it is pressed into its approximate original shape and allowed to dry. It is then that the fit with the ball-joint is checked, and it proves to be much closer to the correction that is needed. It is also important at this time to establish the correct orientation of the ball-joint onto the upper leg, the slot being placed to the front. It helps to compare it to its mate.

Next, a rim of wood putty is applied to the interior of the upper leg, where the ball-joint will be seated into the putty (see below).

The ball-joint is placed into its correct position (see below). By inverting the leg, a chopstick can be used to pack the putty tightly around the seam of the two parts, inside the leg.

The photo below shows the two legs, with the repair leg on the left, as the putty is allowed to dry.

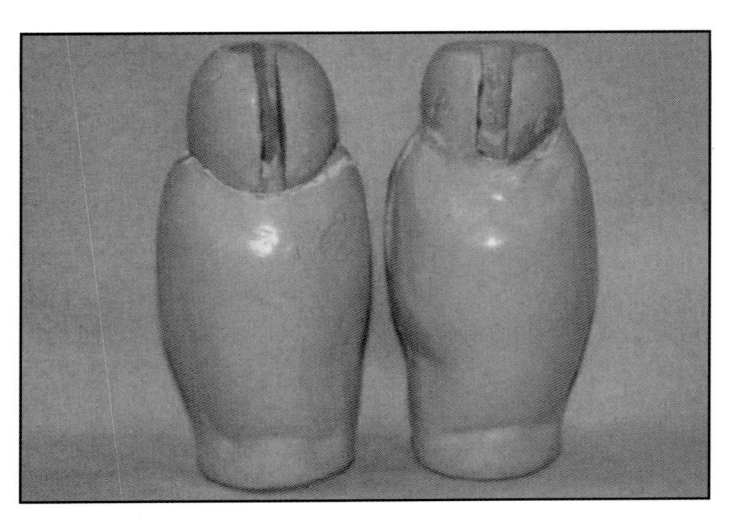

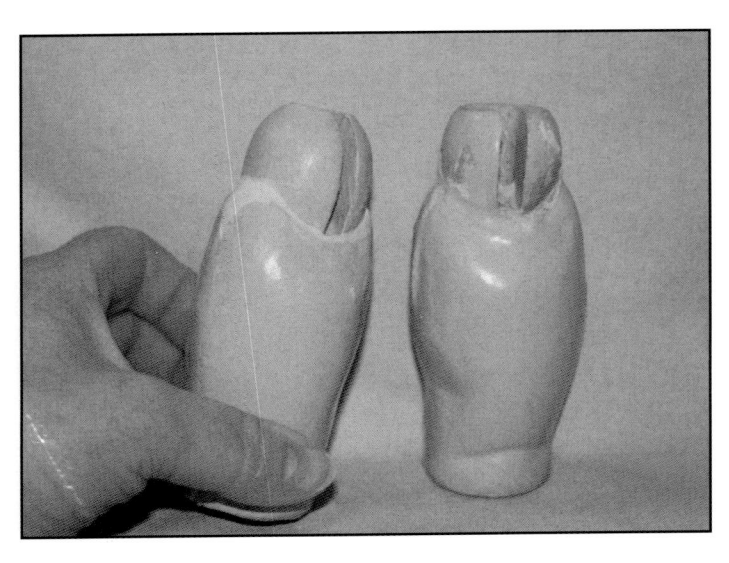

The finished work is then painted, washed, and varnished. Repaired leg on left, un-retouched mate on right.

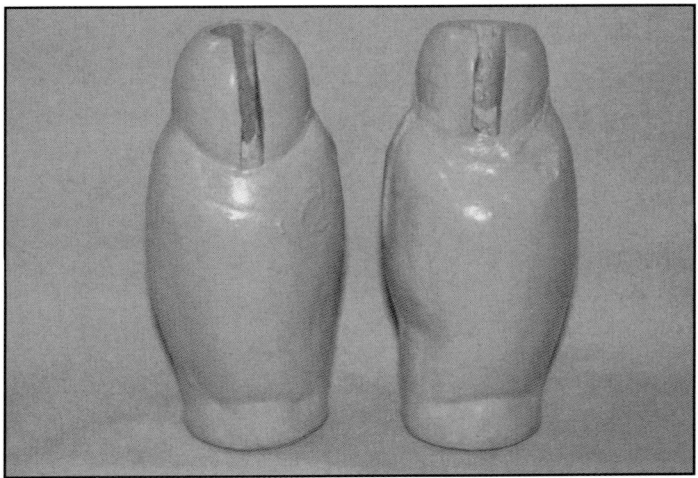

Now that the parts are firmly held together by the interior putty work, the outer seam line is also filled and allowed to dry.

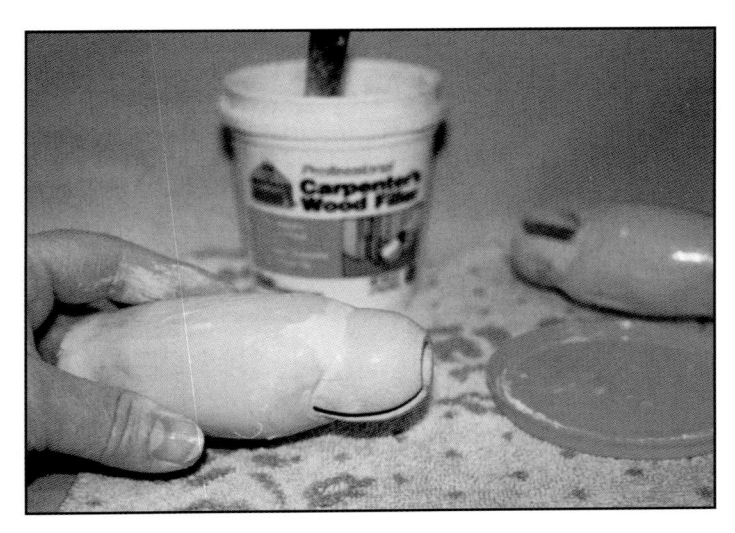

The "proof of the pudding" is a photograph of our little orthopedic patient, standing firm and level on her two feet!

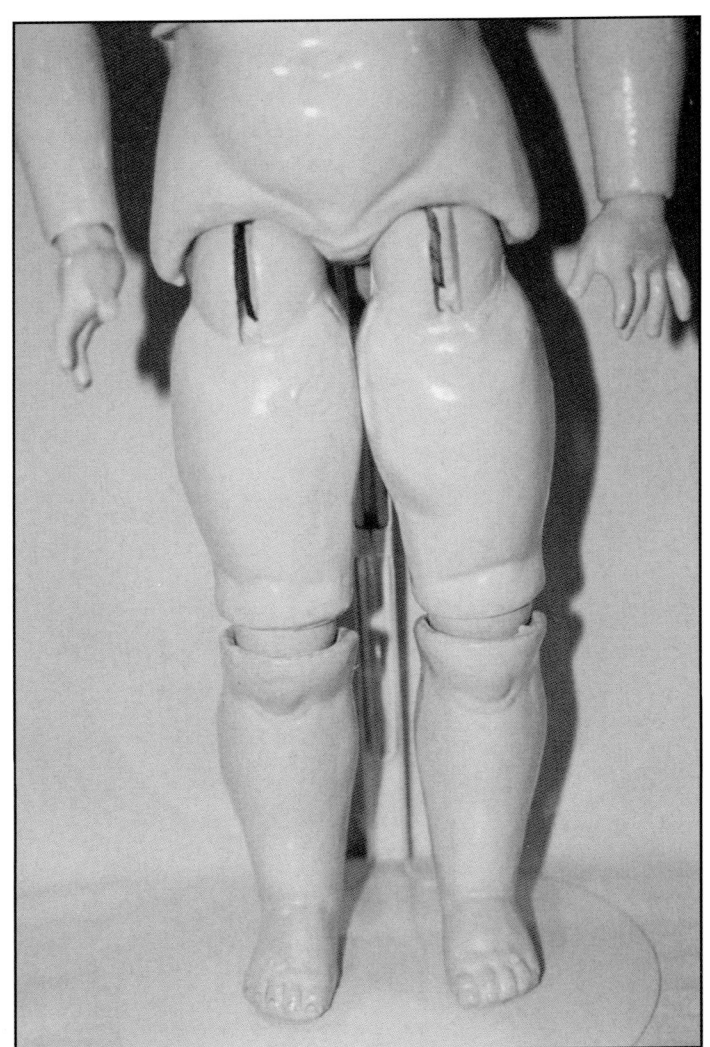

Next, the outer seam line is thoroughly sanded.

Replacement of Missing Hook Inside a Leg

Another type of small repair to a leg is the replacement of the wire hook at the knee. The sketch below shows the leg interior with the wood disc added and the cotter pin shaped into a loop to accept a hook and stringing cord. The wood disc and cotter pin, prepared as drawn, can be puttied into the top of the leg. A hemostat (or a long hook fashioned from wire) is attached to the assembly so that it doesn't fall down into the leg during the work.

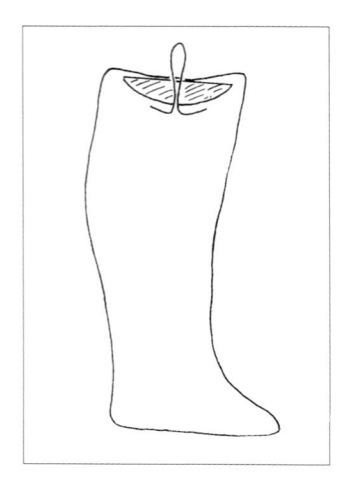

Re-setting a Displaced Socket

Our fourth repair task for this section is a relatively simple, out of body armhole socket. After checking for proper orientation of the socket into the armhole, a rim of wood putty is applied on the inside of the armhole, where the socket will be positioned (use the same techniques as for the upper arm ball-joint repair, above).

The socket is put into place into the wet putty. Holding it in place with one hand, tamp the putty into the seam with the other hand, using a chopstick passed into the opposite armhole opening.

Allow this putty application time to dry.

Next, apply wood putty to the outer seam line, and allow to dry.

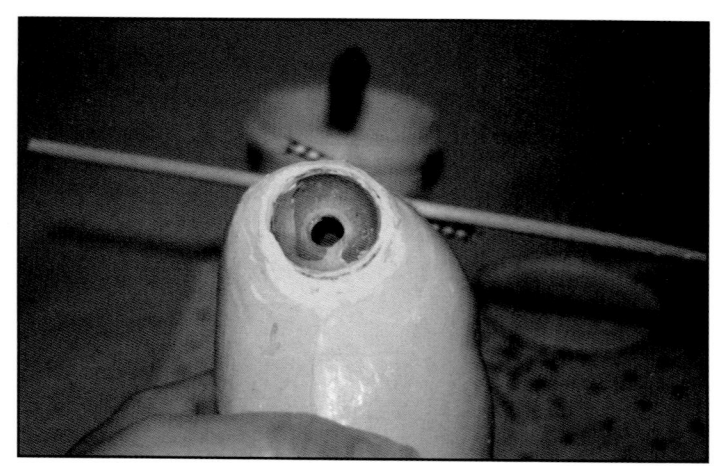

The putty is thoroughly sanded when dry.

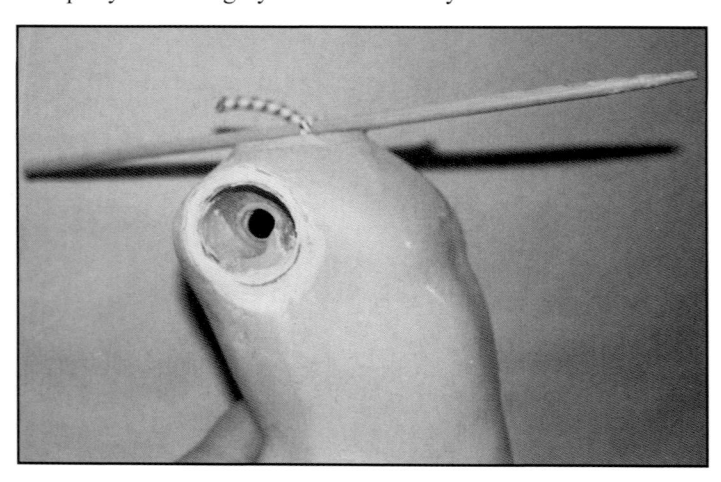

Here, the socket area has been painted, washed, and varnished; when dry, the arms may now be strung.

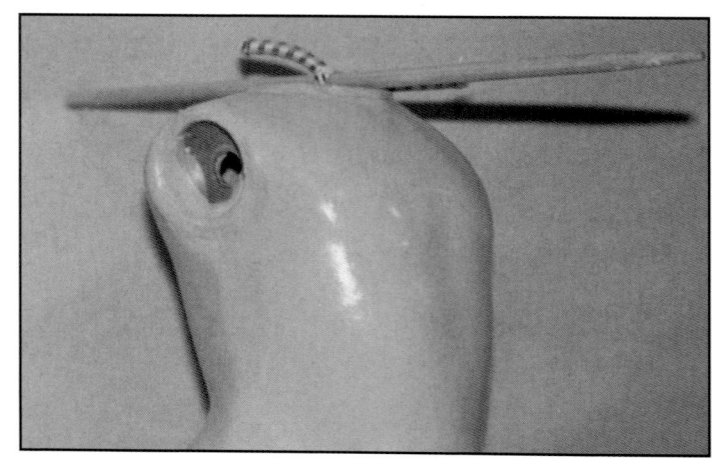

Re-setting a Neck Socket

Our fifth repair task for this section is the re-setting of a neck socket.

As shown below, the neck socket is not only loose, but damaged, and with little support in the body opening. The doll head, which had been strung with very tight springs, had been pulled into the body cavity with little neck showing above the neckline, as seen on the following page. This definitely causes problems in appearance and function as well.

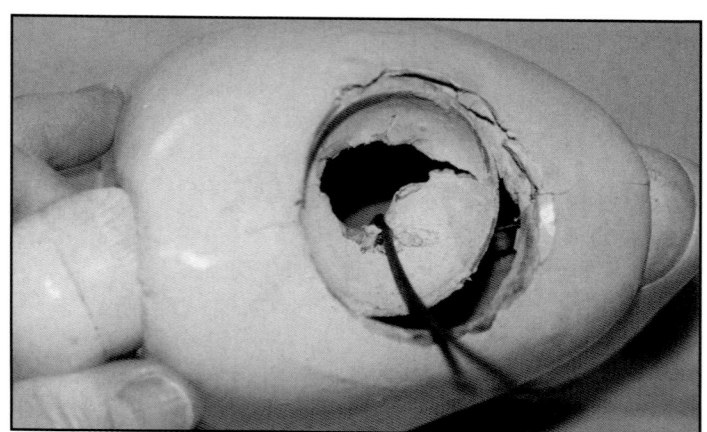

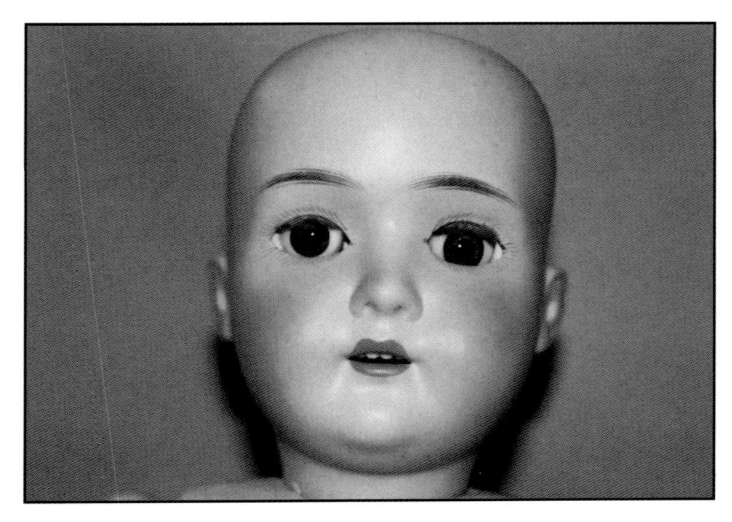

The socket is removed from the body, marking the alignment of both pieces by pencil.

A reminder: thorough cleaning of all parts would be done at this time, before beginning repairs.

A rim of wood putty is placed around the inner rim of the neck opening. The socket is fitted into the putty and aligned as the pencil marks indicate.

Packing the putty into the seam line on the inside of the body is more difficult to access, as the distance from the leg openings is longer. A long piece of wood dowel rod helps with this task.

When dry, putty is then applied to the outside socket rim, smoothed with a wet finger or paint brush, and allowed to dry. (see below).

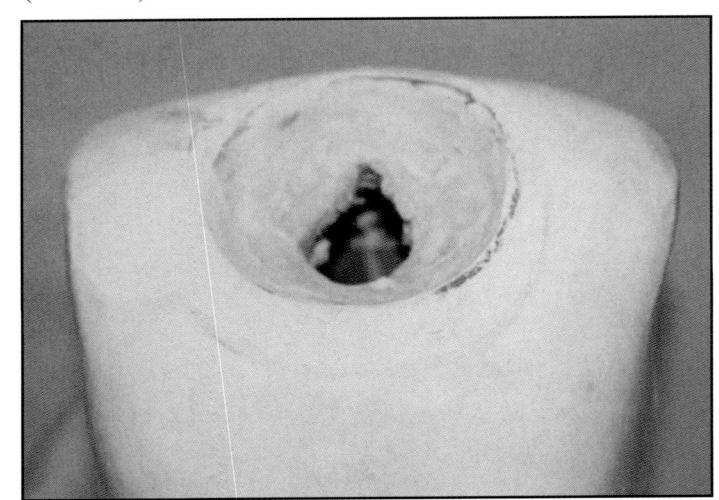

The putty repair is sanded, painted, washed, and varnished; drying time is allowed between each coat.

The finished neck socket repair is almost invisible! (See above right photo.) How much better she looks with her head properly positioned and her neck visible above the neck socket. You will "see" this doll in later pages; her name is "My Girlie". Watch for her, she's a cute one.

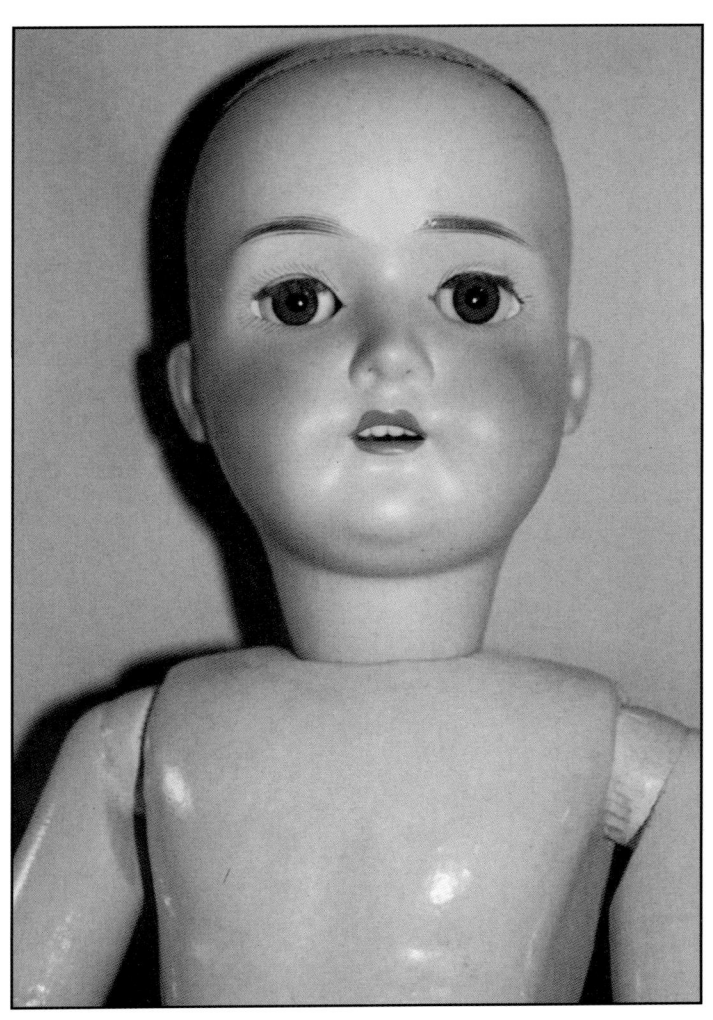

Replacement Fingers

The putty used for body repairs is also of use in replacing missing fingers on composition and ball-jointed bodies.

Drill a 1/16th inch hole into the hand where the finger is missing (see below and following page).

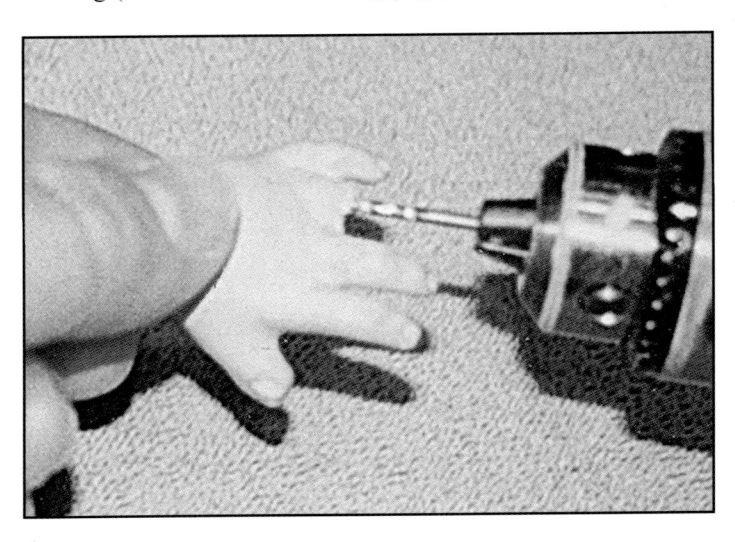

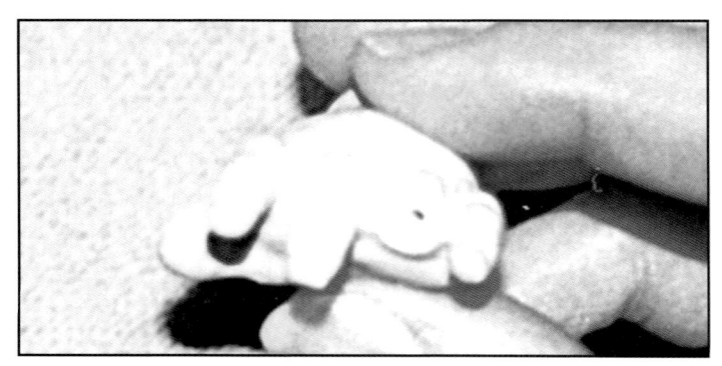

Glue in a piece of wire slightly shorter than the desired finger length.

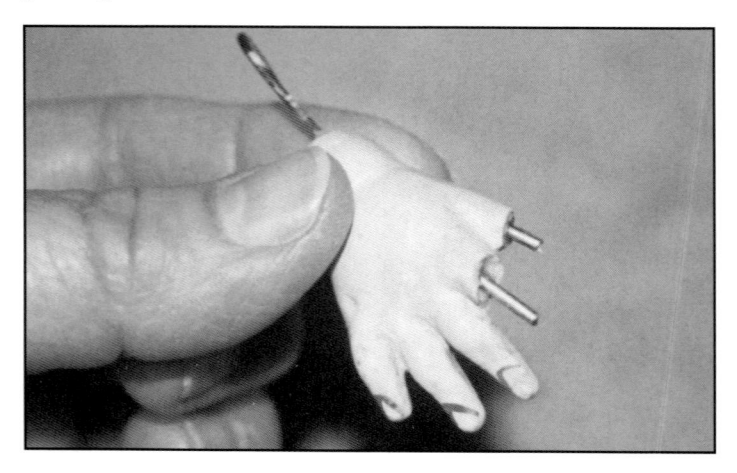

When the glue has dried, form the finger shape about the wire "in the rough" with putty using a small spatula or moistened paint brush.

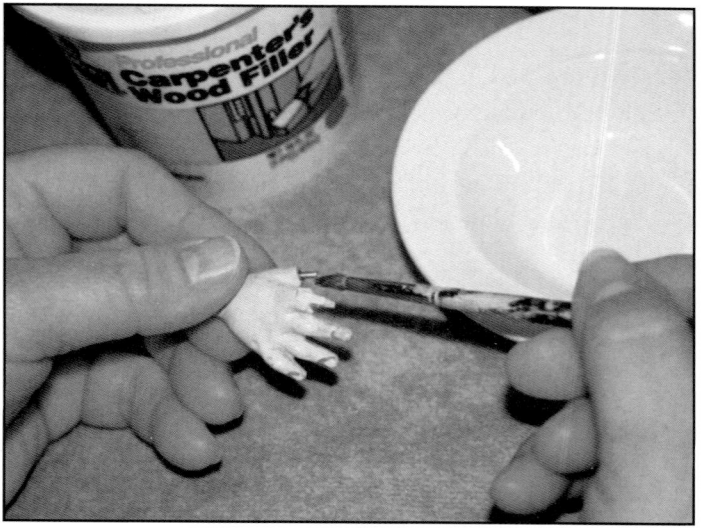

Adapt or create some type of apparatus from which to hang parts such as this hand that need drying time (see above right photo). A miniature quilt rack makes a handy drying rack and doesn't take up too much space in the work area. One or two items can even be hung from kitchen cabinet door handles, if your kitchen isn't too heavily trafficked by others.

In this first application of the putty, the beginning shape is a rough version of the finger; use less material than needed, rather than too much. Sanding putty to reduce the size is tedious, so apply a little less than you think is needed, rather than more.

When the first modeling attempt is dry, go over it with a critical eye. Add a little putty where needed and smooth with a wet paint brush. Again, hang to dry. When this second modeled shape is dry, sand down any areas which need to be removed.

Then administer a final coat to give the final shape to the finger, by using a cream-thickness mixture of water putty (see below). This powder, mixed with water, can be brushed on with a soft paint brush, loading on where needed to give the proper shape. This coating can be completed within minutes, and little or no sanding will be needed.

Again, hang the hand to dry, then paint to match the rest of the hand using acrylics, as above. To more closely match the original paint color, it helps to add a wash after the first paint coat is dry. The purpose of the wash is to mimic the shading that results from the aging of shellac coats that were often used. Mix a drop of a strong yellow color with acrylic paint thinner, coat the area, and blot or wipe off. Allow this application to dry and then apply a coat of acrylic varnish.

Finger (or Toe) Lining

Antique ball-jointed dolls often had what is called "finger lining". These red lines were painted on, sometimes across the knuckles of the doll fingers, sometimes between the fingers in the "web" of the hand, and sometimes on both areas. In the above photo, the little finger was missing and has been replaced. How odd it would look if it were left as it is in this photograph, without the finger lining that the other fingers have. The next photo shows the completion of the finger lining on the replacement finger. Finger lining is added by using an (doll) eyelash brush and thinned red paint. Some restorers use a fine-line red marker for this purpose. This is acceptable if the ink is permanent, as water-base marker can blur with the acrylic paint or can be rubbed off. The photo below shows the finished fingers (from page 38), which match the original fingers so well, it's difficult to spot the repair. Success!

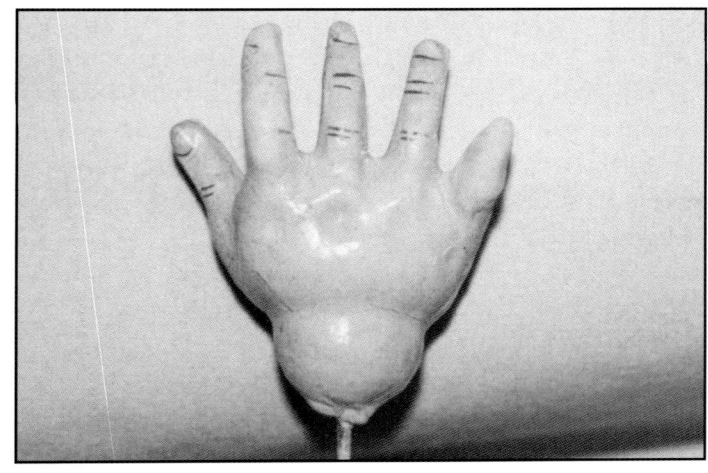

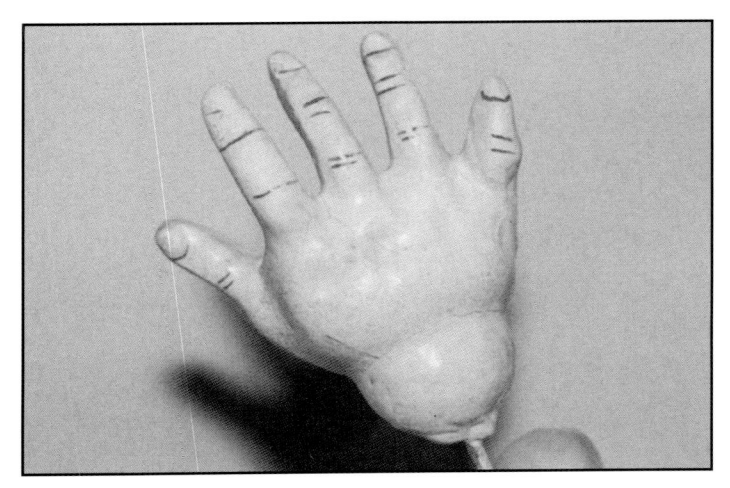

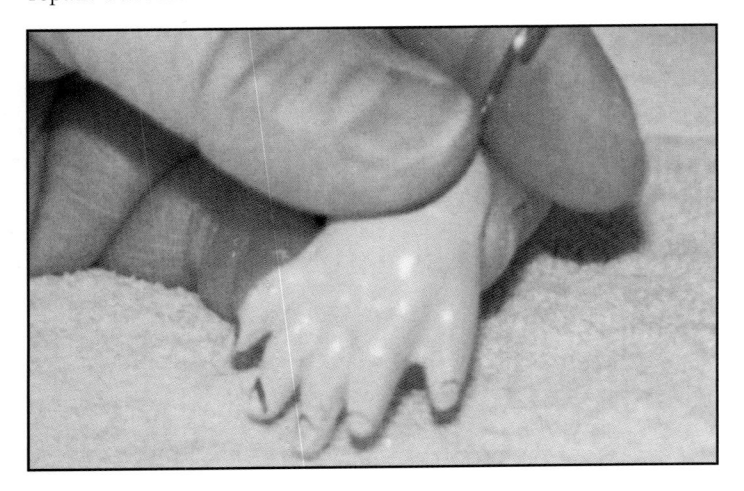

Re-stringing Antique and Antique Reproduction Dolls

Supplies Needed for Doll Stringing

Simple but essential tools are needed for re-stringing dolls:
Two pair needle nose pliers.
A selection of surgical clamps (hemostats).
"S" hooks in a variety of sizes.
A stringing hook.
Various sizes of stringing cord.
Smooth dowel rods or chopsticks.
High quality elastic loops of various sizes (cut from medical grade rubber tubing).
Stringing supplies and tools include both the common variety of "S" hooks, but also heavier ones available at hardware stores. Hardwood "wheels" are ideal for neck buttons and stronger than the type usually sold by doll supply companies.

Some General "Stringing" Rules

Most antique dolls should be strung "head-last". There will always be exceptions, primarily dolls with five-piece bodies. The main reason to do the head last is the obvious concern about possible breakage or loosening of sleep eyes. Regardless of your level of experience, it's easier and safer to string the body, then have a helper assist you with the head. If you must work alone, the chopsticks and hemostats really become your friends!

Be sure to work on a soft padded area to prevent scuffing to the bisque head, while handling.

Stringing dolls is probably one of the most difficult things to teach, as there will be so many variations of techniques to achieve the best result. While the antique ball-jointed bodies are quite similar, simple variances, such as unusually small hooks planted in the hands or small holes through which stringing cord must pass can pose particular problems.

The goal here is to give good basic instructions, mention common pitfalls, and trust the learner to devise his or her way out of unique obstacles.

Basically, the first step is to choose a stringing cord with a stretch ratio of two to one. That is, a six-inch piece that can be stretched to twelve inches. The need for this much elasticity is that a little less accuracy is needed in judging the amount of cord to use to achieve the right amount of tension.

Stringing Ball-Jointed Child Bodies

While ball jointed bodies have more parts than other types, they are actually easier (usually) to string
Usually these bodies have fifteen pieces;
a torso
two upper arms
two elbow ball-joints
two lower arms
two hands
two upper legs
two knee ball-joints
two lower legs
Some will have elbow and knee joints as part of the arm or leg and have less pieces. Others may have extra ball joints at the shoulders and hips and have as many as 19 pieces.

As you disassemble the body, take note of the sizes of ball-joints and which are elbow and which are knee. If you are worried that you will confuse these pieces, disassemble half of the limbs and leave the others together as a work example.

This is the time to clean these pieces, as explained in the cleaning section. If you are tempted to skip this cleaning, just try a little cleaner and see how much soil is present. Even if the body isn't very soiled, the cleaning also does a remarkable job of conditioning the paint and polishing up the finish.

Make sure that you also clean all dirt, old stringing cord, stray hooks, the occasional dead bug (!), and any other "left-overs" out of the body torso, and dust out the inside of it as well.

Before Stringing is Begun

Doll stringing is difficult to teach because there will always be a doll body that comes along with a different structure or nuance of design that changes some aspect of how stringing can be accomplished. We can only hope to present to you most of the pitfalls you might encounter if you are new to this type of work.

Orienting Stringing Cord Around Body Braces

Sometimes there are wood braces inside the torso in the chest area.

They sketch below is of a torso interior, showing a brace from back to front. When you string a body with a brace positioned in this way, the neck hook must not be so long that when the stringing cord is installed and the head attached, the neck hook might "sit" on this brace, interfering with a snug stringing. While you will not encounter a brace in this front to rear configuration very often, you need to be aware of the possibility.

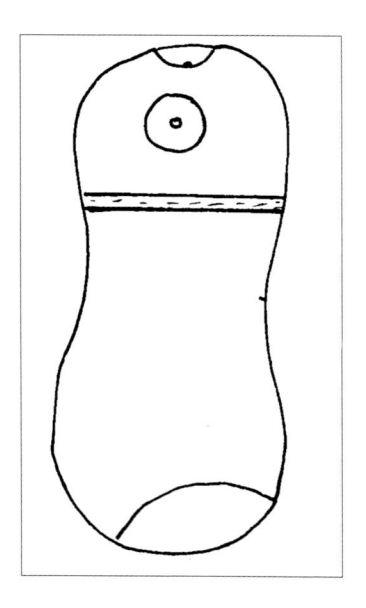

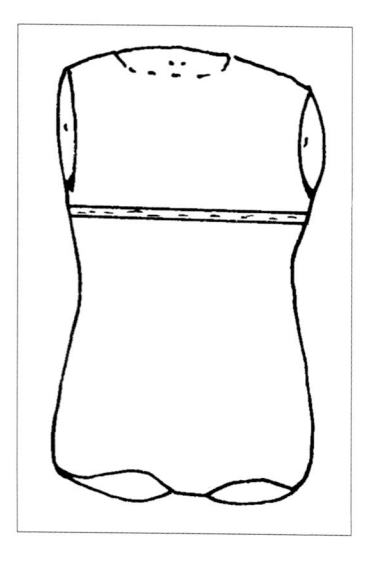

A second, less problematic task in stringing dolls is the other type of brace that you might encounter. The sketch below is of a torso interior with a brace positioned from side to side in the chest area, usually just under the arm sockets. This is a much more common type of body brace to encounter. When you string a doll with this side to side type of body brace, you must orient the leg to head stringing so that the stringing cord from both legs passes either behind or in front of the brace. You cannot string the doll with one leg-to-head cord in front of the brace and the other behind it. If you do this, when the head is added, the cords will be straddling the brace and may press down on it. If this occurs, either the head will seem loose because the elastic cords cannot pull down all the way, or the brace will be popped out of position by the pressure. Choose to string either behind or in front of this type of brace, depending on which side has more space. There will usually be more room behind the side to side brace than in front of it.

You must also consider how you will pass the arm stringing cord through the torso chest. After you have strung the legs and before you attach the head, string the arms, orienting the cord through the body so that it does not pass through, but rather behind or in front of, the leg stringing cords.

Other Little Problems

Sometimes the doll that is being strung has very small hooks planted in the hands. It will be difficult to crimp a stringing cord knot onto a too-small hook. The simplest solution is to add another hook to each hand, on the original hook. Use one just large enough to accomplish the task.

Another "small" problem occurs when the holes through the upper legs are too small to accommodate the stringing hook. This usually occurs when the upper leg has a wood ball joint at the top, rather than being made totally of cardboard. The solution for this problem is to string the legs in reverse. Begin with the stringing cord at the top of the upper leg and pass it down through the hole, through the ball joint, place a knot at the end, and attach the knot to the hook of the lower leg, crimping it tightly. Now the leg is strung, and the cord can be pulled up through the torso as usual.

A Stringing Diagram for Ball-Jointed Bodies

The sketch below is a stringing diagram for a basic German ball-jointed body.

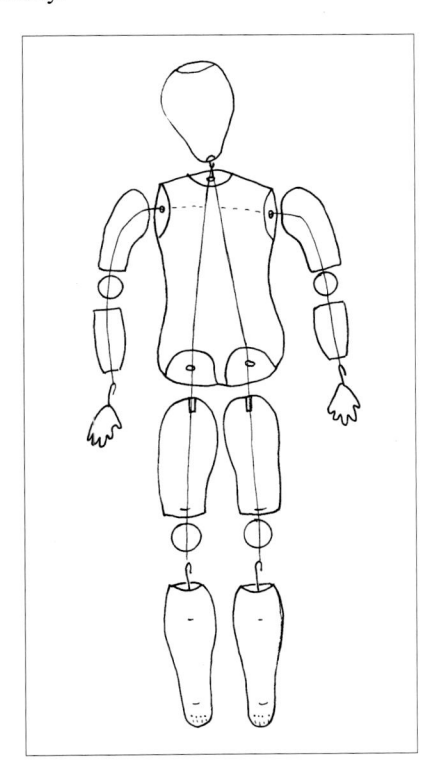

Stringing legs to body

A stringing hook is a tool on the order of a long crochet needle, with more flexibility. Stringing tools are available, or one can be fashioned from a stiff wire clothes hanger. Tape off the cut ends to avoid nasty scratches to person or doll! Attach cording through hooks in lower legs, through upper legs, into torso and out through neck.

Check to make sure that the cord is not twisted inside the body and that, if there is a wood support inside the body, both cords go either behind or in front of it, not one behind and one in front. It is greatly helpful to have a partner, when stringing dolls for the moral support if not the needed third and fourth hands! Tie off one side at a time, using a round dowel rod (chopsticks work well for this) to tie against. This type of rod will roll with the movement of the cord and, with normal care, will not scratch the doll's surface. Tie snugly, but not overly tight. Check leg tensions; these should be approximately the same.

Gentlemen who do doll repair sometimes tend to under-estimate their strength, and may string dolls too tightly. This is especially critical if there are hairlines in the doll head. Unnecessary pressure is not advisable!

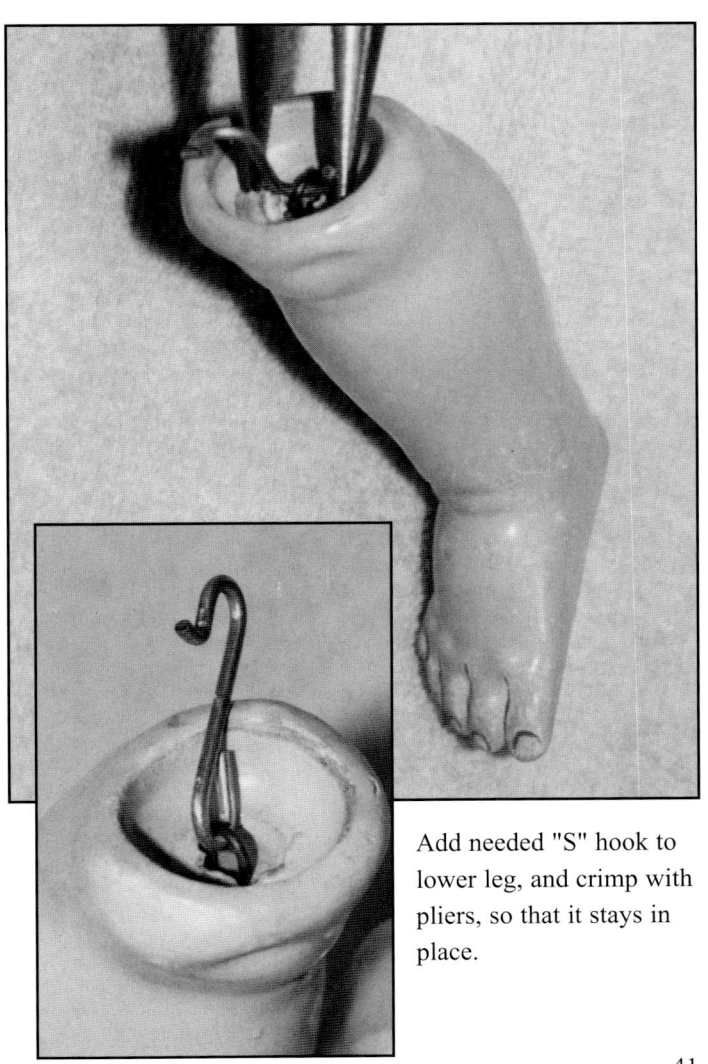

Add needed "S" hook to lower leg, and crimp with pliers, so that it stays in place.

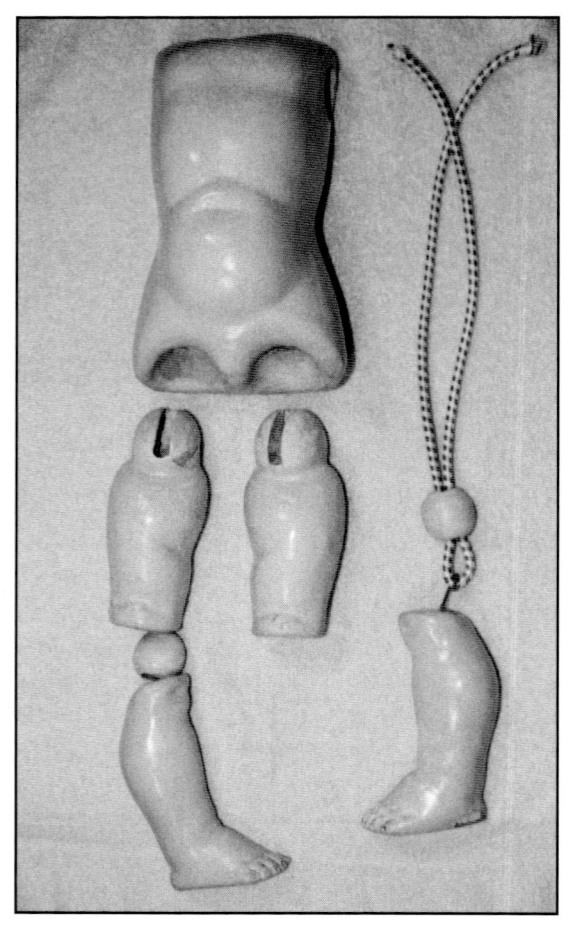

String cord through "S" hook, add knee ball-joint.

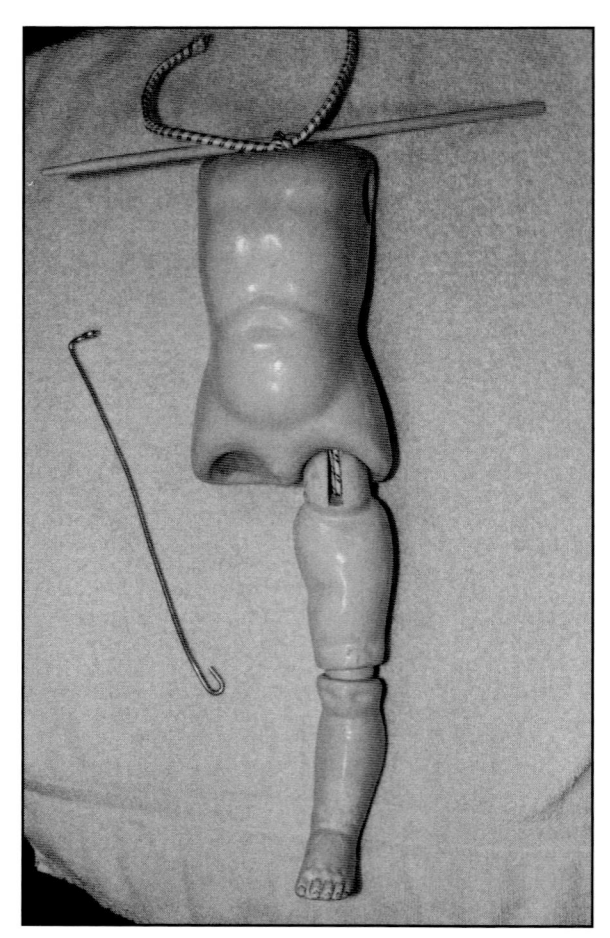

String cord through torso and, pulling to stretch it tight, tie it, double knotted, around a dowel rod at the neck opening.

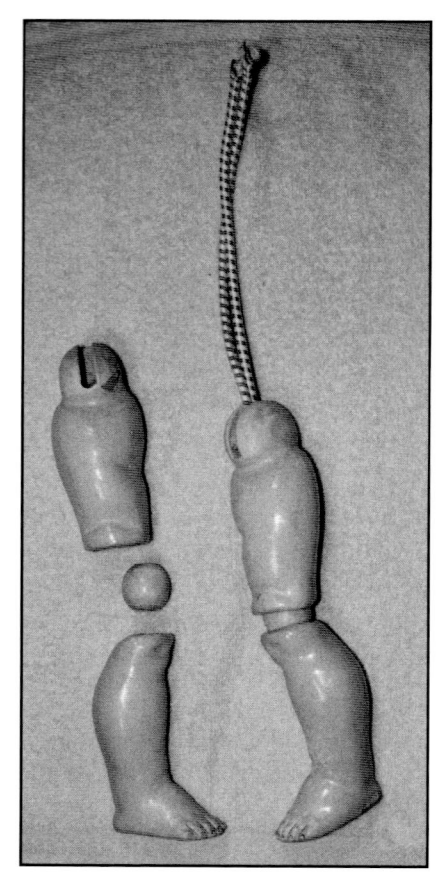

Stringing Ball Jointed Arms and Hands

Begin by placing a knot at the end of the stringing cord. Attach the hand and crimp the hook about the knot so that it is tightly held by the hook. To the cord add the lower arm, ball joint, and upper arm (see below).

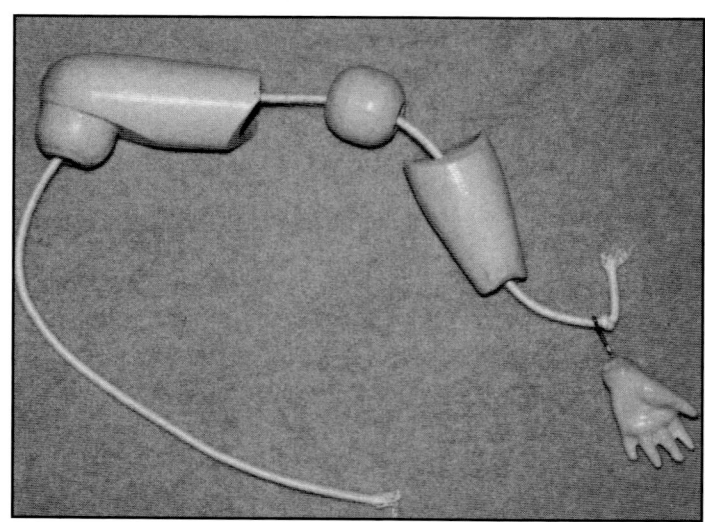

String cord through upper leg,

Then draw the cord through the doll's chest, using a stringing hook as needed (make sure that you are adding the arm to the correct side of the body. Nothing is more aggravating than stringing parts and realizing that you have put them on backwards. Dolls do not like this!) Now reverse the order of the arm pieces, first upper arm, then ball joint, then lower arm, and lastly the other hand.

In the photo below, the remaining hand is still having fingers built, so we are using a clamp to hold the arm stringing until the hand is finished and the stringing can be completed.

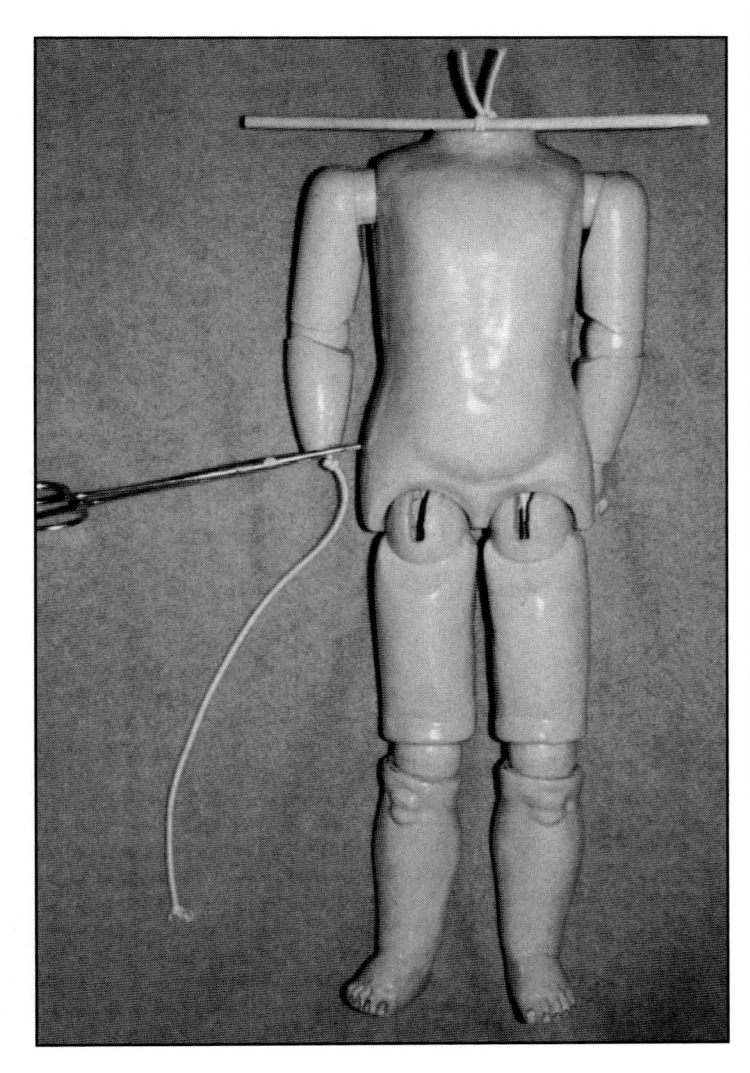

In the following photos the fingers are finished and the hand can be added (a chopstick makes it easier to get the hook into position) while a helpful friend prepares to remove the chopstick.

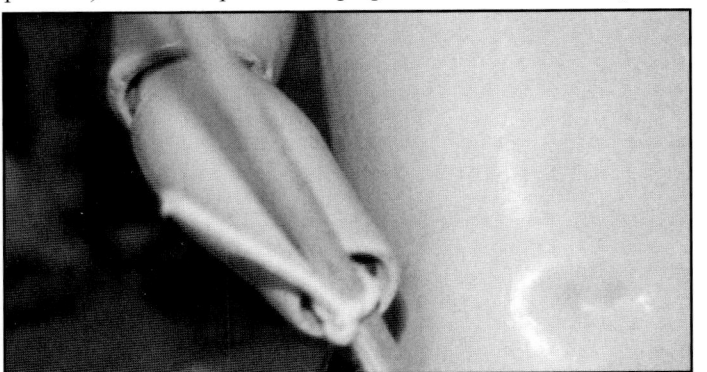

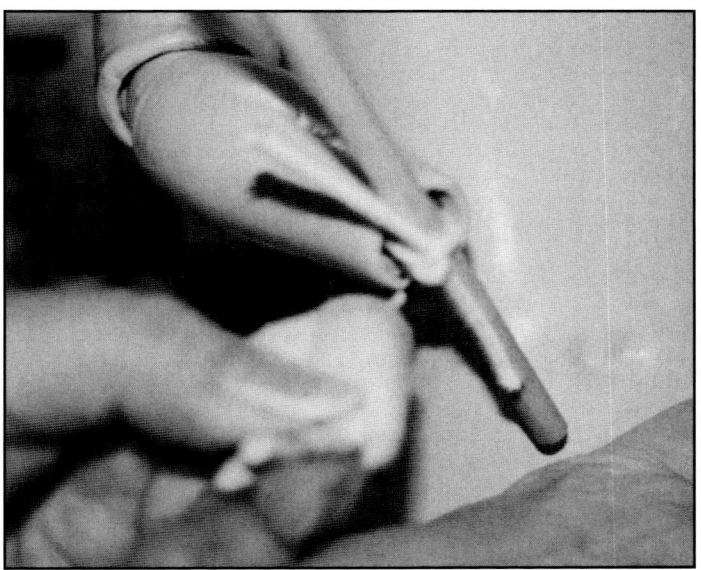

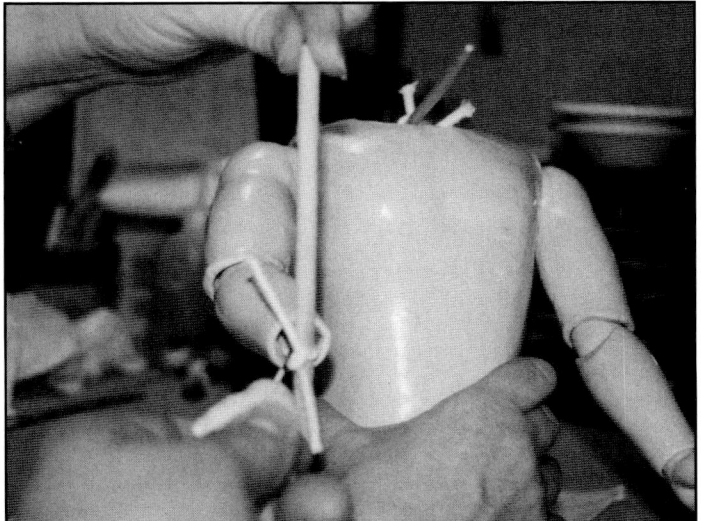

Making a (Better) Neck Button

These instructions for neck buttons will provide you with a neck button that is superior to any that you can buy from doll supply sources.

The sketch below shows the preparation of a neck button using a hardwood "wheel" and a suitably-sized cotter pin. The tips of the cotter pin are each bent using a pair of needle nose pli-

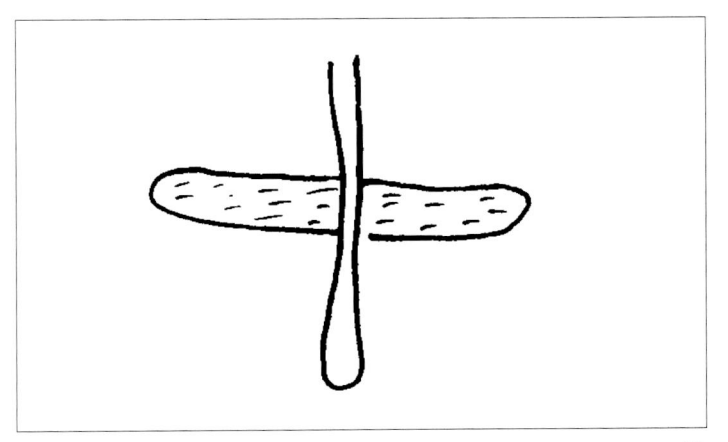

43

ers so that they are flat against the button and leaving enough of the looped end of the cotter pin exposed to attach the neck hook.

The following sketch shows the completion of the neck button, with the attachment of the neck hook. In actual stringing, it is easier to attach the neck hook to the cord at the body neck socket and then slip the head hook through it. It requires less maneuvering of the (oh, so breakable) bisque head.

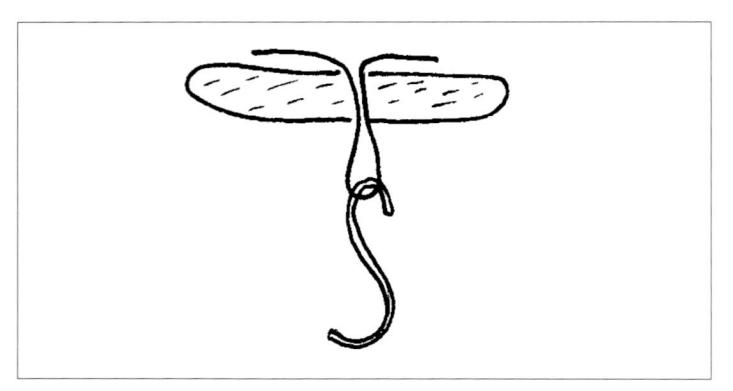

Attaching the Head

Attaching the head is left until last when stringing antique bisque dolls. It is helpful to have the help of a second person. If such help is unavailable, one person can string the head by holding the doll down upon the workspace with forearm and elbow. This is not recommended for beginners.

Prior to attaching the head, do a safety check of the leg-to-neck stringing. Make sure hooks are securely crimped about the elastic and that knots are tight and not slipping.

Neck buttons are another crucial link in holding the antique head, in the stringing configuration (mentioned earlier, the hooks in the lower legs are the other half of this link).

A neck button which is weak or too thin may crack, and must be replaced. If the original neck button is present and in good condition, of course it may be reused. If you wish to use the original neck button check the condition of the wire hook to be sure that it is not corroded and ready to snap. Replace this wire whenever you are in doubt about its lasting strength.

The unpainted hardwood wheels sold by craft supply stores are an excellent choice for neck buttons. They have smooth, rounded edges, are available in several sizes, and are an adequate thickness, which guards against cracking under the stress of stringing.

A variety of sizes of cotter pins, available in hardware stores, are the other item needed to prepare a neck button. Cotter pins are much stronger than the usual hook sold by doll supply companies for neck button attachment.

To make a neck button, first choose a size of wood wheel which fits easily inside the doll's neck, large enough not to fall through the neck opening, but not so large that is doesn't "seat" at the bottom of the neck. It should not ride up on the inside wall of the neck.

Using two needle nose pliers, place a cotter pin through the neck button hole, holding the cotter pin with one pliers and turning down the longer tip of the cotter pin with the other pliers.

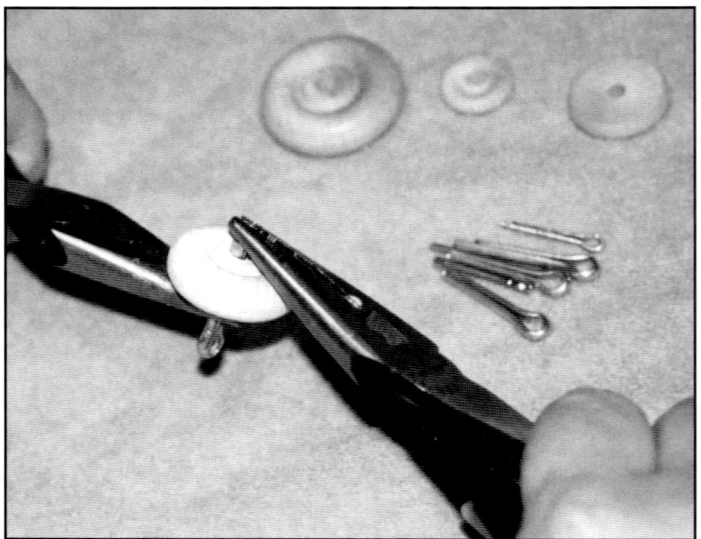

Again, using two pliers, turn down the second tip of the cotter pin (see below). Add a suitable size of "S" hook to the loop of the cotter pin to check for ease of movement in the loop. This S hook will be moved to the cord at the neck of the doll body, as follows:

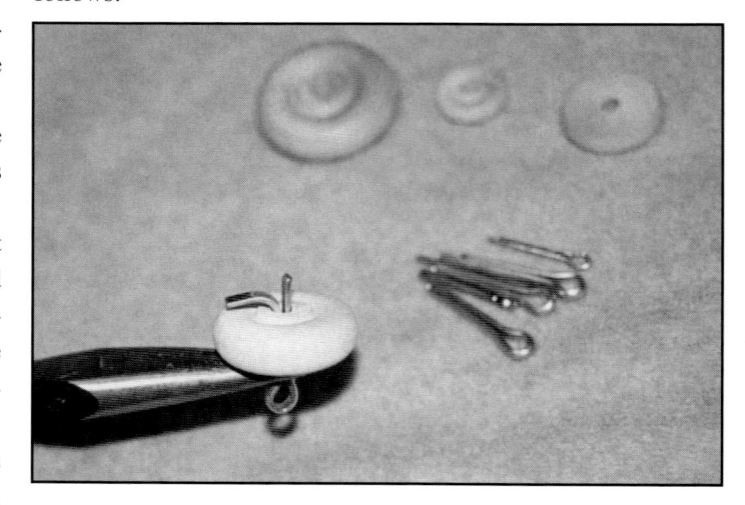

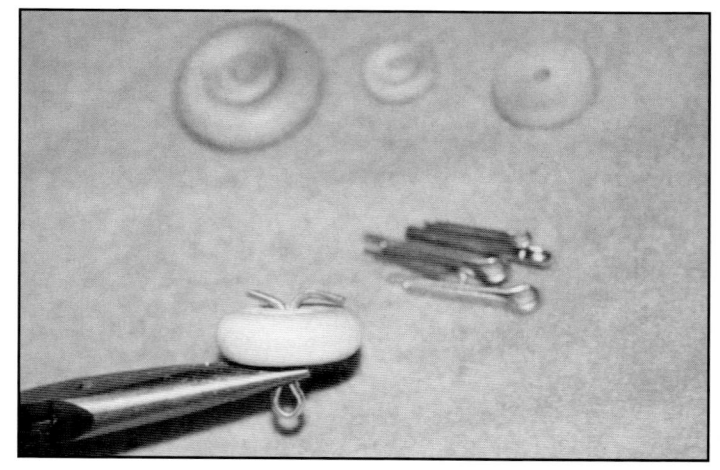

Add an "S" hook, chosen in the previous step, to the stringing cords at the neck opening of the body.

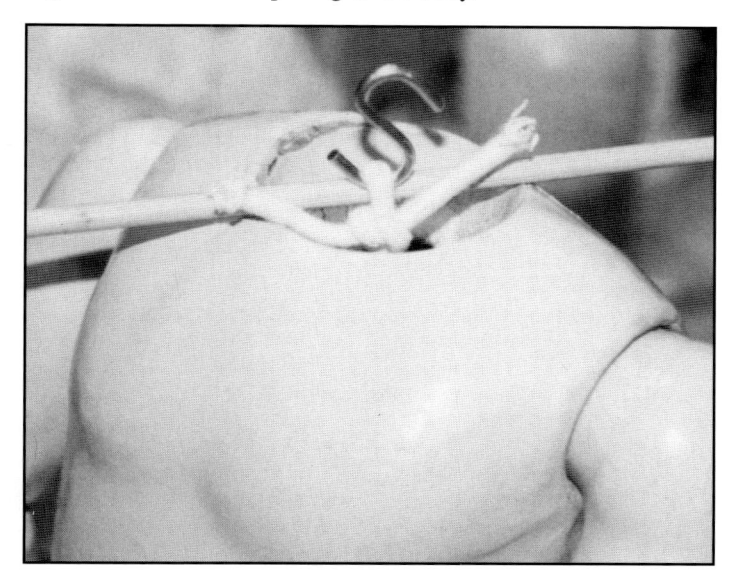

Use a hemostat (scissors-shaped clamp) to hold this hook in place if it seems that it will fall into the body. You will have your hands fully occupied elsewhere!

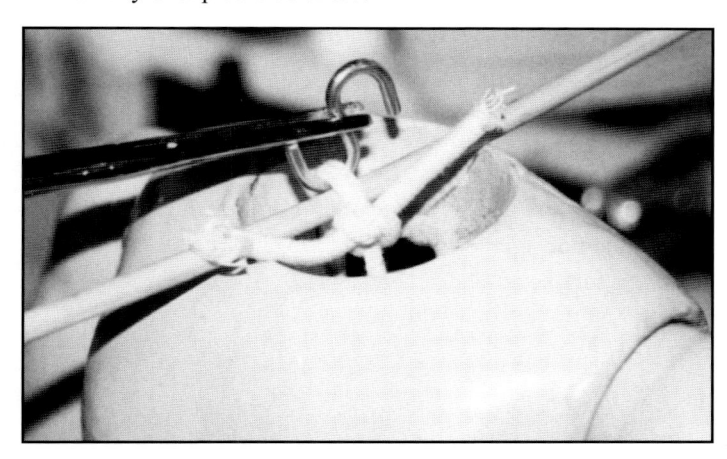

Slip the head onto the hook at the body neck opening and, holding the head firmly with one hand, remove the dowel rod which has held the leg stringing with your other hand. Leave a couple of fingers between the head and the neck opening of the body for cushioning. Having someone stand by to help the first time you attach a bisque head is a comfort if not a necessity. Tuck the loose ends of the stringing cord into the neck opening and ease the head into place. If you have stuffed the head to keep the eyes motionless, that material may be removed now.

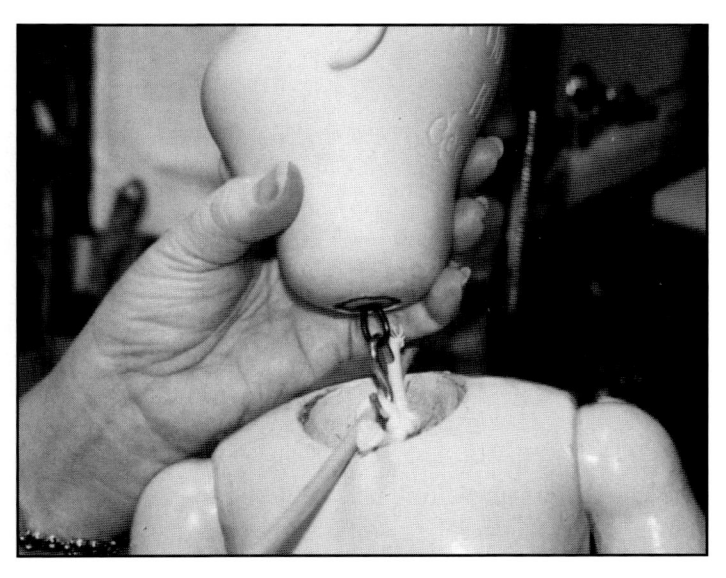

Stringing Five-Piece Body Antique Dolls

Stringing Bodies with "Staked" Limbs

Quite different from the higher quality ball-jointed bodies, re-stringing the five-piece, sometimes called five-piece-crude body (because some were not well finished) requires a very different procedure.

To complicate the work, sometimes these dolls had no hooks in the arms or legs, but rather the stringing cord was simply "staked" into a hole in the limb, glued in place, and then slivers of wood, or "stakes," were forced in around the cord to take up the space and hold the cord firmly in place. The outer surface was filled over with some type of putty and painted.

The sketch below shows the method used when the five-piece body was originally strung. The head was merely hooked to either the arm cord or the leg cord. Heads that are attached in this way usually will be cocked a little to one side or the other, as they are not squarely anchored by this stringing method.

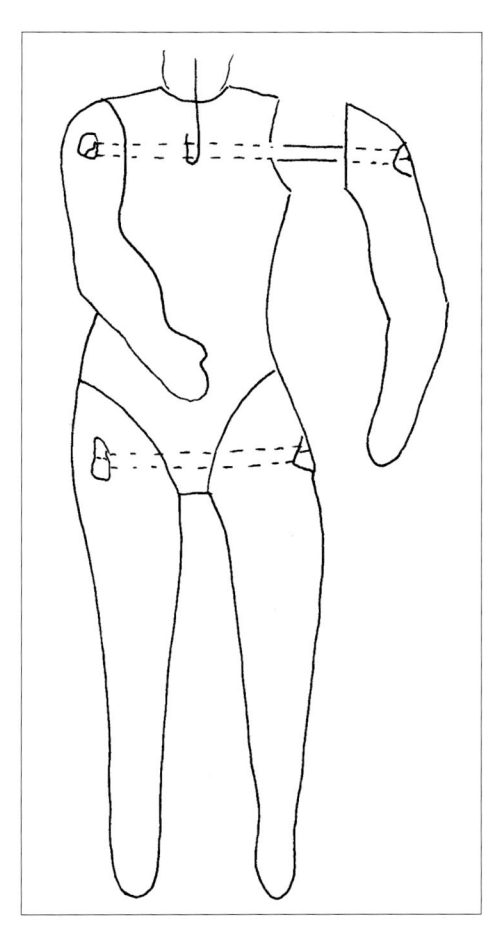

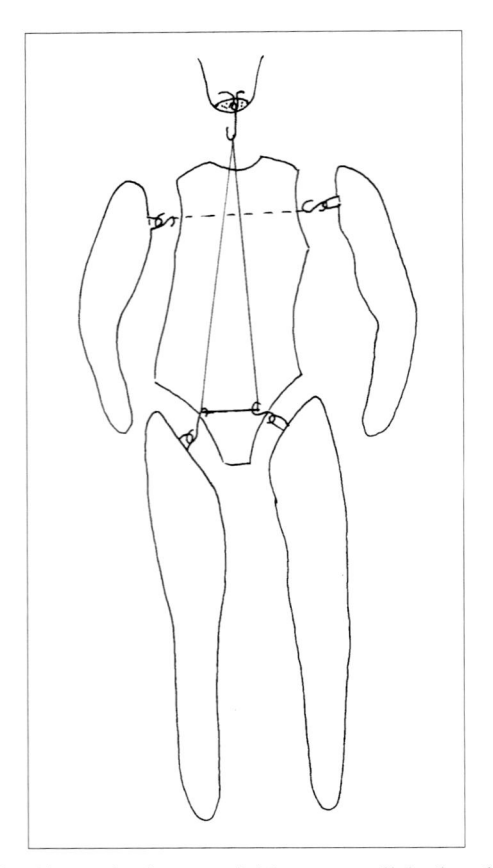

While this method was quickly accomplished and therefore economical in a doll industry which was trying to survive difficult economic conditions, it creates problems when re-stringing is needed.

There seems to be no reason to repeat the original method, because when the doll needs re-stringing a third time, in years to come, the same problems will still be present. Rather, it seems logical to improve the method so that the next stringing needed will be simple to accomplish.

The improvement needed is simple-to-install wire loops into the limbs so that the arms and legs may be strung using "S" hooks and stringing cord. The head will string to the legs in a triangular configuration as shown in the sketch above.

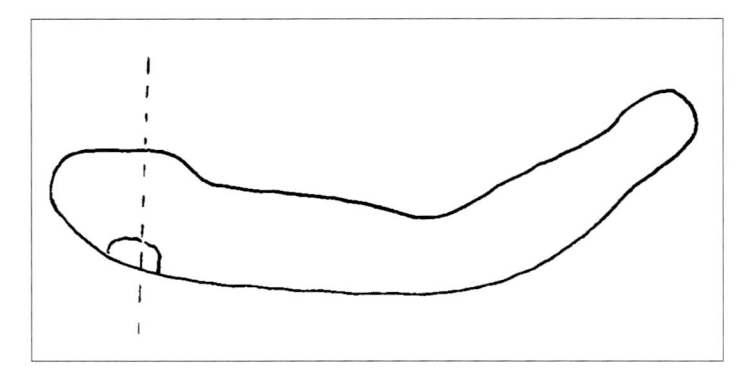

First, drill through the arm (or leg) to remove the stakes and old stringing as in the sketch above. It's wise to start by drilling a small "pilot" hole, using a 1/16-inch drill bit, and then use a larger bit to finish the job.

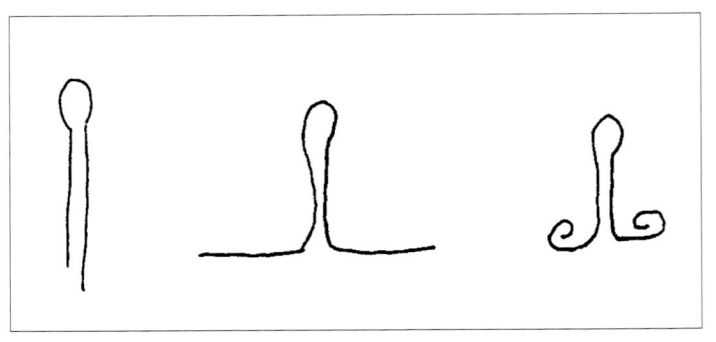

Using an appropriately sized cotter pin, or very stiff wire, craft a coiled loop to insert into the limb, as shown above.

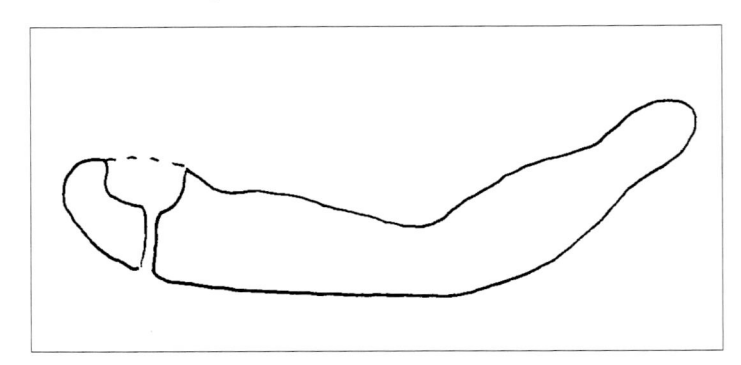

The stringing hole on the inner side of the limb will likely need to be enlarged to receive this coiled cotter pin, as shown above.

The sketch below shows a sketch showing the installation of the cotter pin into the limb. The cotter pin is set in with wood putty, first on the inner side and allowed to dry. Then the filler to the smaller hole on the outer side of the limb is added. When dry, the outer surface is sanded and painted.

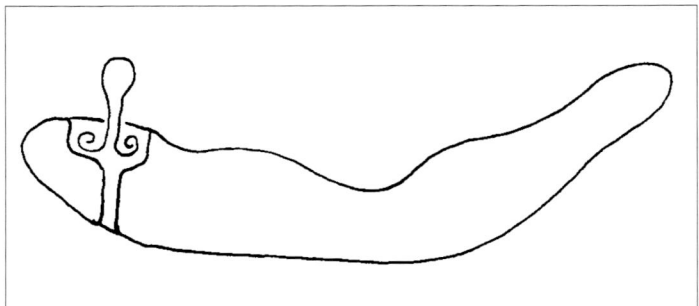

When all four limbs are finished, the doll may be re-strung. While this procedure is not quick to accomplish, it is far superior to the original method.

Setting Hooks into Glued-On Ball Joints

Another type of doll body which lacks hooks where they are needed is the body which has ball-joints glued on to the limbs, usually on both the arms and legs. The stringing cord was originally installed as shown in the sketch on the following page. A small nail was placed through the stringing cord. The cord was then passed through the ball joint and the entire assembly glued into the limb, usually with lots of what was probably animal hide

46

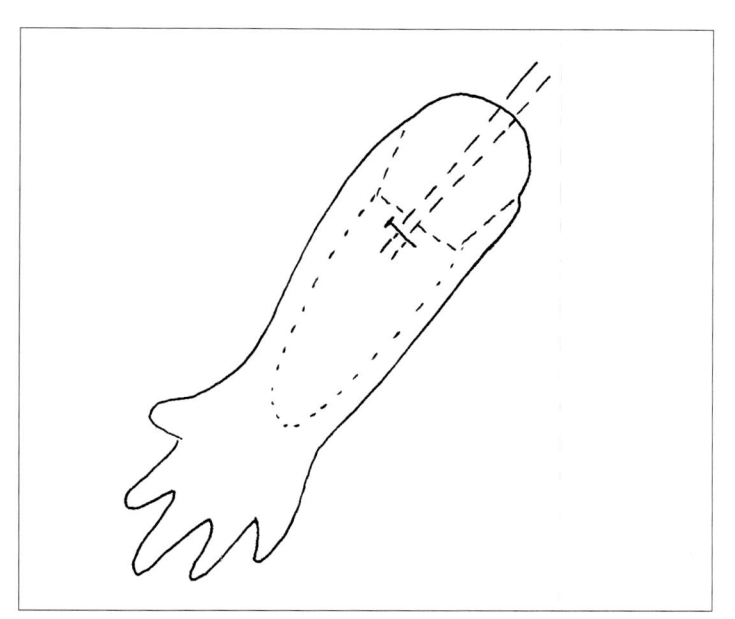

glue. The joint was puttied into place and the entire limb painted. The cord was then passed through the body and the entire procedure reversed to attach the limb's mate.

This method does not seem to be simple in its original conception, and it is curious why it was used. It needs to be improved upon, so that when the doll will need re-stringing, it will be less difficult to accomplish.

The goal, as in all stringing tasks, is to have hooks available on each limb to which the stringing cord may be attached.

Two methods will be presented, one which is better and one which is easier.

The first method requires that the ball-joint be removed from the limb to gain access. If the joint is already out of the limb, or loose, this method is preferred.

If the ball-joint is firmly in place in the limb and you don't wish to potentially damage the limb by having to dislodge it, you will want to use the second method, below.

Continuing with the first method, clean off the glue from the ball-joint and remove the old stringing cord. Choose two washers, one almost as large as the bottom surface of the ball-joint. The second washer should be smaller, but not small enough to pass through the hole of the first washer.

The sketch below shows the assembly of the washer combination. Using a piece of gauge 22 steel wire, prepare a long "S" hook by looping through and crimping onto the smaller washer. Then pass the wire through the larger washer and then through the ball joint from the underside. This wire will need to be long enough so that it loops above the ball-joint, where a knotted stringing cord is attached.

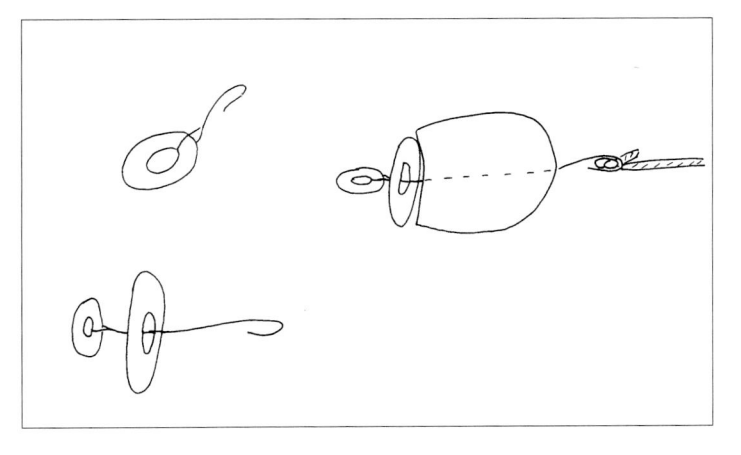

The following sketch which shows the ball-joint and stringing assembly glued back into the limb using wood glue. After sufficient drying time for the glue, the seam around the ball joint will need to be filled with putty and painted.

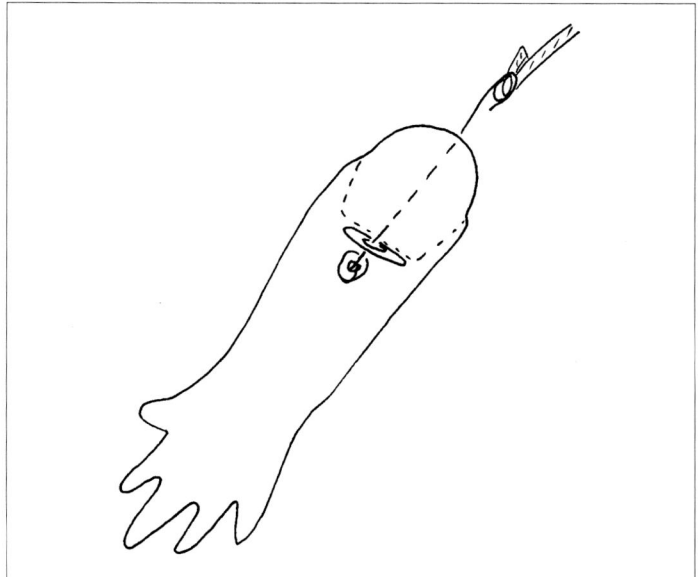

When the first limb is finished the stringing cord is then passed through the body, pulled tight, and clamped off with a hemostat. Here the procedure is reversed to add the limb's mate. The cord is knotted and trimmed, and wire for the S hook is crimped on. This wire is then passed through the prepared ball joint and the first and second washer are added (the hemostat is left in place, as you need to be able to do this work with no tension on the end of the stringing cord. If you remove the clamp, the stringing cord will retract into the body, taking your unfinished "S" wire with it!) The ball-joint is now glued into the limb using wood glue and allowed to dry. The seam line is then filled with putty, allowed to dry, sanded smooth, and painted.

This method, as with other types of doll repair, has several stages which require drying time for glue, putty, or paint between each stage. It is much more time consuming than the second method, which follows. It does give the glow of accomplishment to the restorer, however, especially if the doll is a significant one, and you want to do the very best for him or her!

In the second method for glued-on ball-joints, the goal again is to place hooks in the limbs so that this and future re-stringings may be accomplished easily.

The first sketch on the following page shows the same type of glued in place ball-joint that was discussed in method one. The old stringing cord and glue needs to be drilled out, as indicated by the long dotted line, going through the slotted hole in the ball joint. Use a drill bit close to the size of the stringing hole, probably at least 3/16-inch.

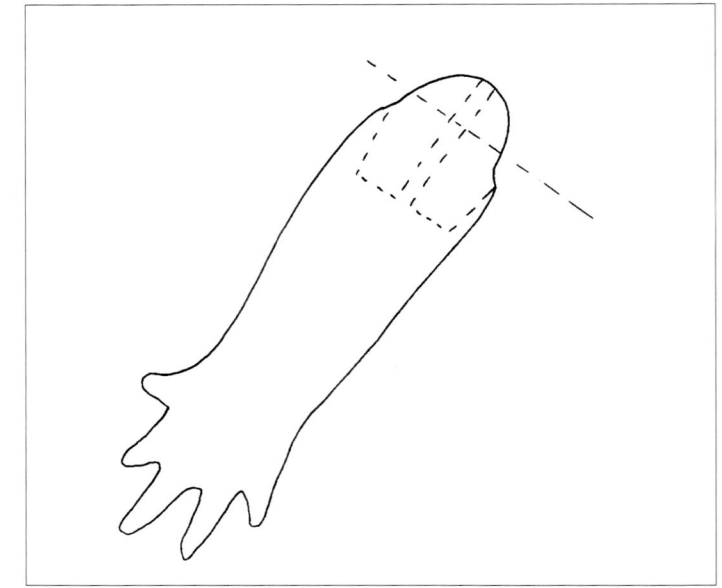

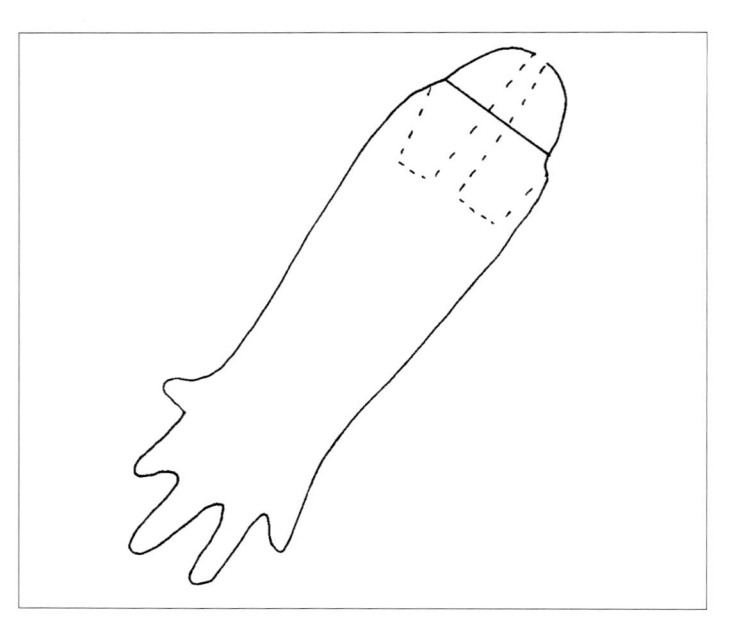

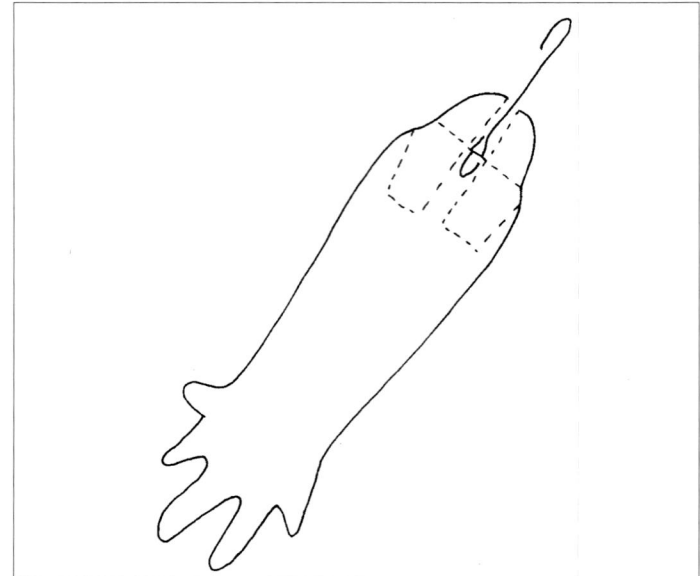

The sketch at the top right shows another drilling line, this one perpendicular to the other and passing across the joint in a "T" pattern. This hole is being drilled to accept a "pin" (use a 1/16-inch drill bit) which will be a finishing nail of proper size. It is onto this pin that you will crimp an "S" hook, so it needs to be high enough in the slotted hole that you can reach it with the hook. (Trim the nail to the length needed by snipping it with a wire cutter or metal snips. It's wise to wear protective goggles while doing this task.

To install the pin, you may need to use a tool called a nail set to tap it through the ball-joint. This is a metal tool which looks somewhat like a pencil, about four inches in length, with a pointed end, for "setting" nails, that is tapping them into place with-

out hammering the dickens out of the surface with a hammer. The hammer taps the top of the nail set instead. In a pinch, you can try to substitute another finishing nail for the nail set tool.

When you have installed the pin, you may need to file off the tips lightly with a metal file, and touch up the tip with paint. Because of their position, they should show only slightly when the doll is strung.

The sketch at the bottom left shows the placement of the pin, indicated by the solid dark line. The dimensions of the sketch indicate the approximate placement needed to be able to attach the S hook easily.

The final sketch shows the completed task. The "S" hook used needs to be long enough so that stringing cord can be attached above the ball joint. If your selection of "S" hooks does not provide the necessary length, you can use stiff (about gauge 22) steel wire to fashion one in the needed length.

Stringing Character Babies

Character babies which have five-piece bodies are usually easier to string head-first. The occasional ball-jointed toddler, a rarer example of baby, may need to be done as you would do a child doll. In the child bodies, the legs each attach to the head separately. If this is the case in a toddler, it will string in the same way a child body is done.

However, the five-piece baby bodies are strung head to legs in a triangular shape. This is because the legs mount into the body with less body shape to hold them in place. If they were strung individually to the head, the legs would tend to pop out of the leg socket area. Stringing head to legs in a triangular path holds the legs not only to the head but to each other, and stabilizes their position in the leg socket. A boring explanation, but a critical concept for success in stringing these cute little guys.

Head First

Our repair subject for the stringing of character babies is clean and ready to be assembled (see photo below).

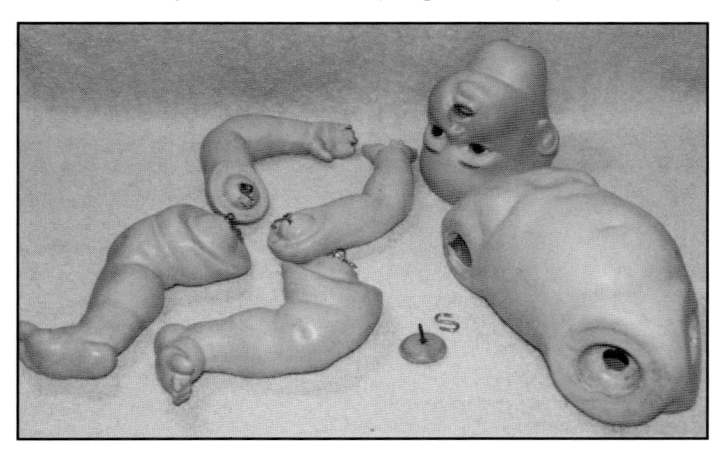

Various sizes of elastic tubing can be cut into loops. Using these loops in combination with various "S" hook sizes gives virtually infinite variability of sizes and strengths for stringing smaller dolls.

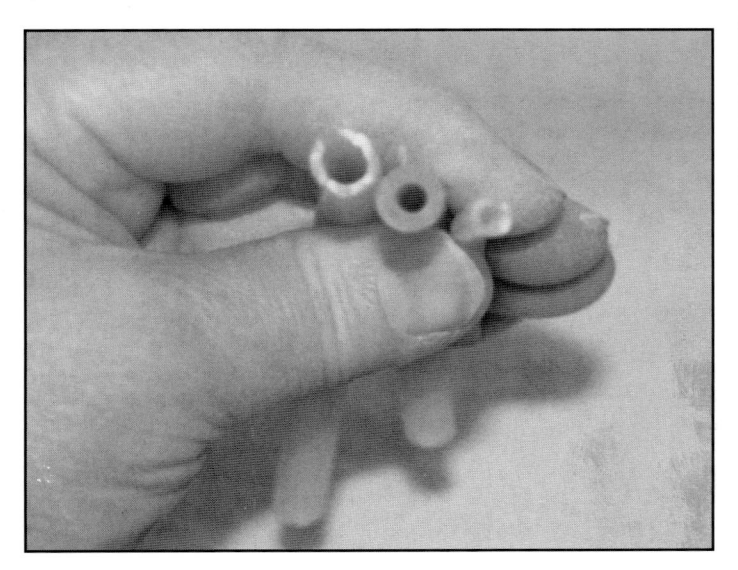

Since you will be attaching the head first, clear your work area of unneeded tools to avoid the risk of bumping into something and damaging the head while you work.

Take note that each limb has an "S" hook; check each one to make sure it is securely planted and will not pull out. If needed, apply wood putty to firm up the wire holding the S hook and allow it to dry overnight before proceeding.

Each of these "S" hooks needs to be crimped onto the elastic loop before you remove the chopstick.

Place the neck button in the neck socket. Below the neck, attach a suitable "S" hook on which you have crimped a series of elastic loops. By stretching these out and comparing to the torso length, you can estimate how many loops/hooks you will need.

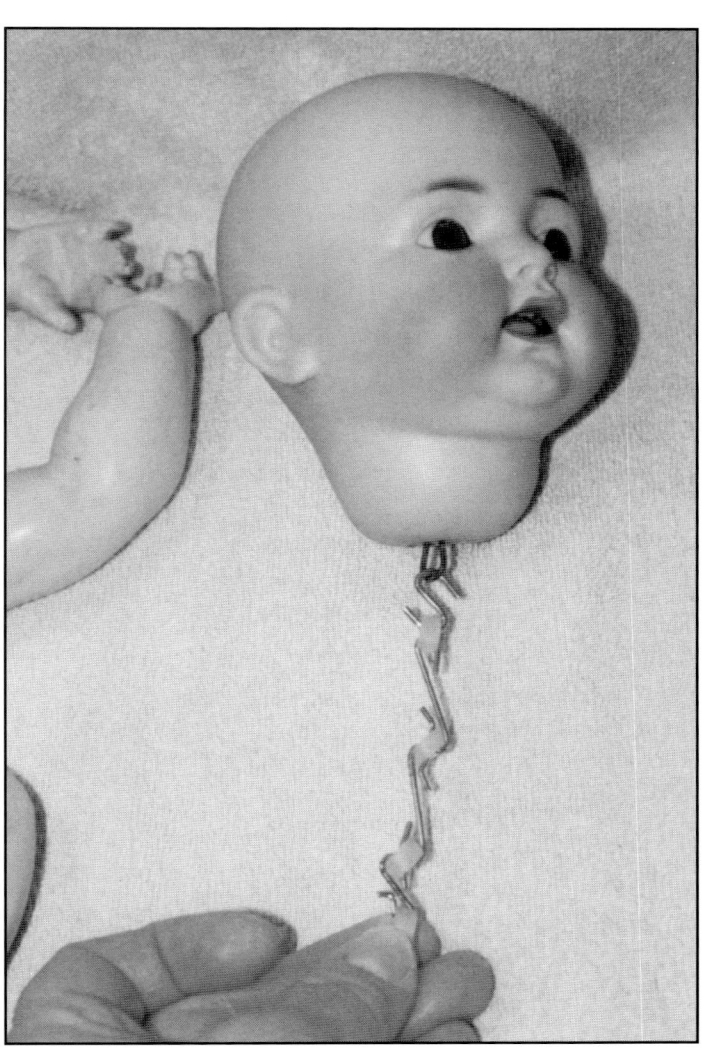

Using the stringing hook, go through one leg opening of the torso, through the neck opening, hook onto the loop section, pull out through the leg opening, and secure by passing a chopstick through the loop. If your loop series is too short to reach the leg opening, go back and add one more hook/loop, or merely change one shorter hook for a longer hook and try again.

Conversely, if your loop series is too long, remove a loop/hook, or change one hook to a shorter hook.

One chopstick on the loop at the first leg opening (see photo below).

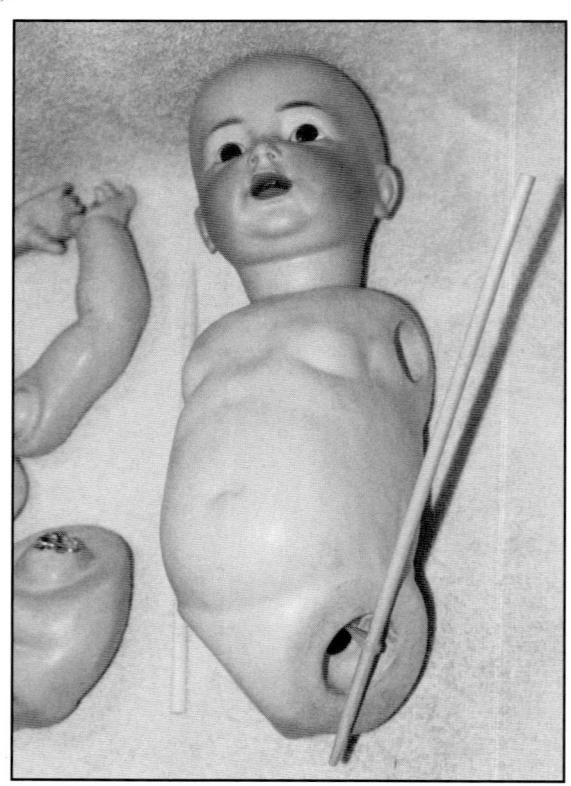

Next, hook on the first leg and crimp. Using the stringing hook, hook through the loop before removing the chopstick. This is easier than fishing around for it inside the body once the chopstick is removed.

Now, pull the loop out through the second leg opening, place chopstick through the loop to hold it in place.

Attach the second leg to the loop. Crimp, and remove the chopstick.

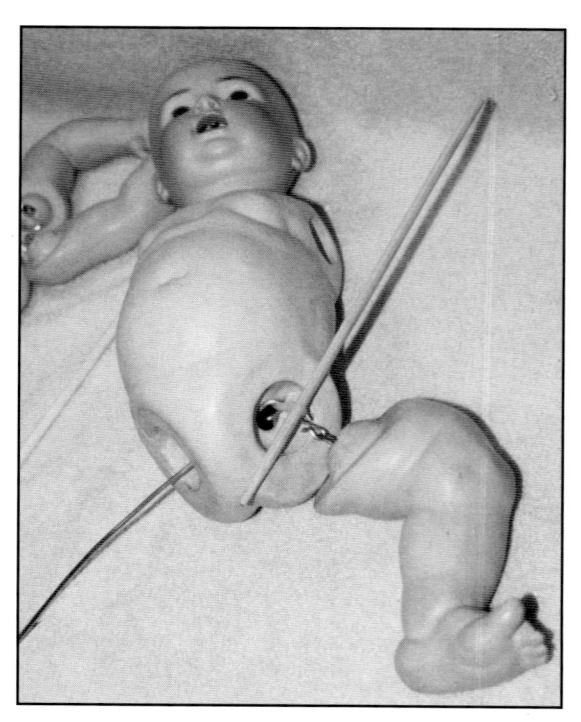

The following photo shows two legs on!

Again, prepare a loop/hook series, stretch to estimate the distance across the chest between the two arms.

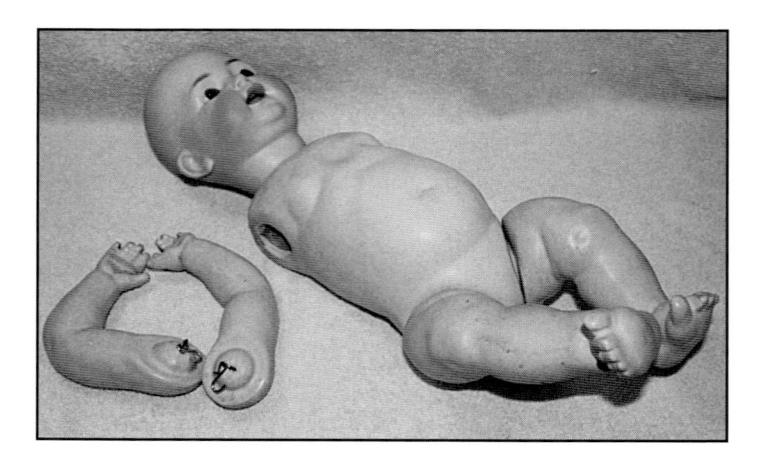

Hook the loop/hook series to one arm, crimp, and pull loop series through the torso chest, using the stringing hook (see below).

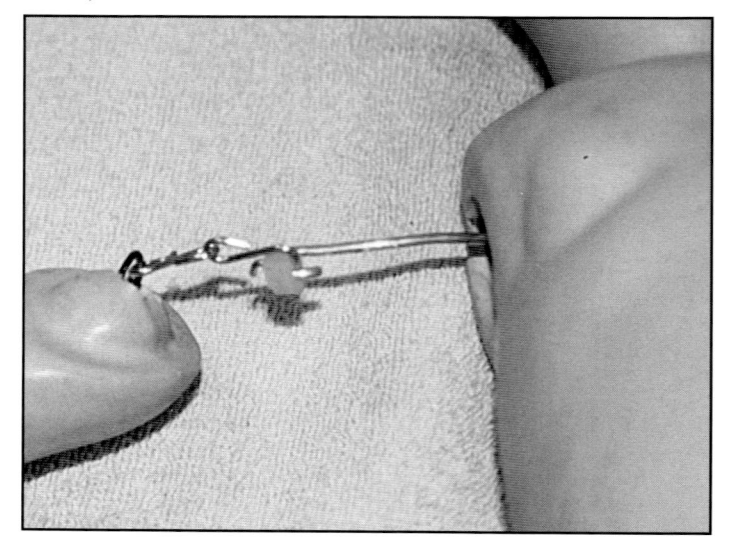

Pull the loop out through the armhole opening and slip a chopstick through the loop to hold it in place.

Hook the second arm onto the loop, crimp, and remove the chopstick.

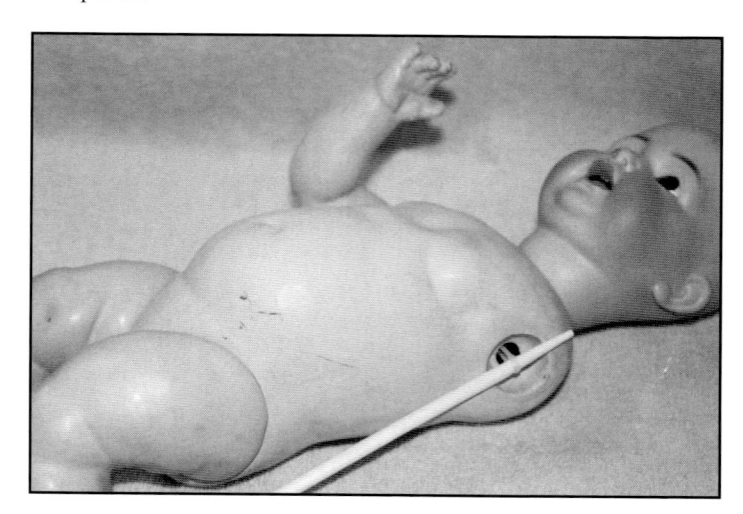

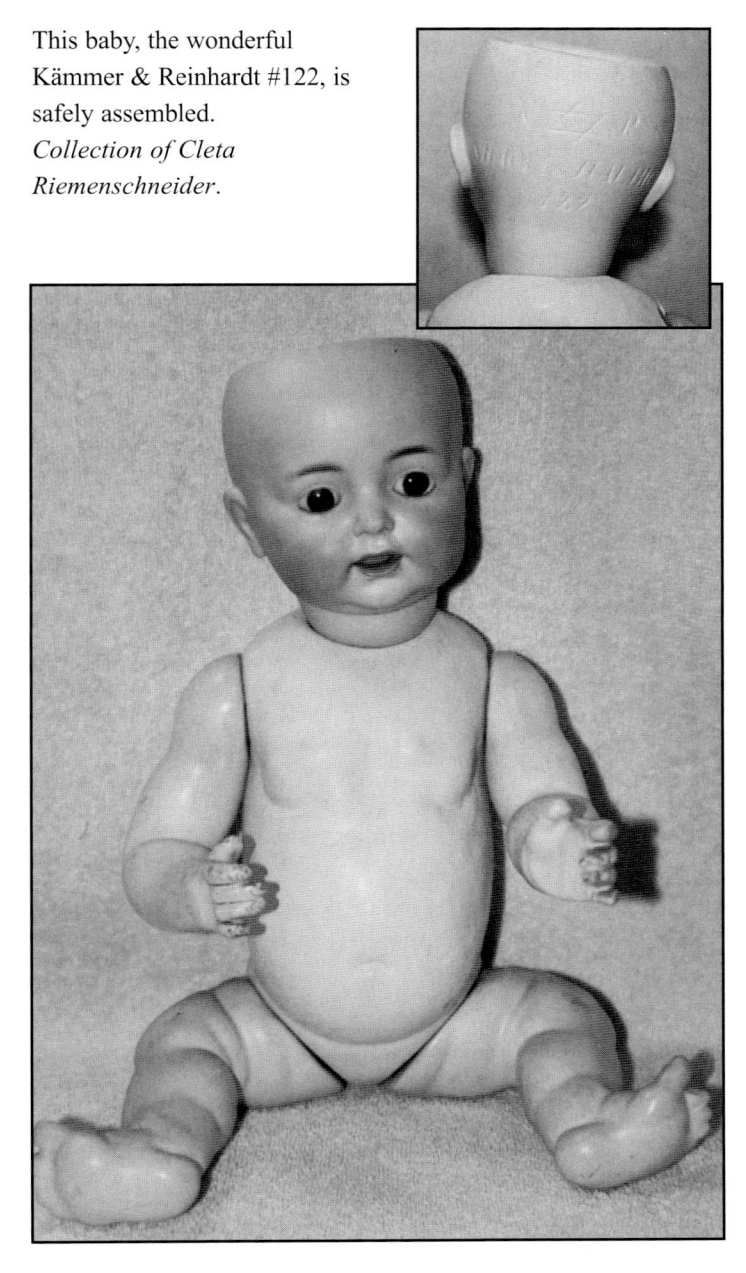

This baby, the wonderful Kämmer & Reinhardt #122, is safely assembled. *Collection of Cleta Riemenschneider.*

Leather and Oilcloth Body Repair

Cleaning Leather and Oilcloth Bodies

These bodies require perhaps more gentleness in handling than do the ball-jointed doll bodies. This is partly due to the condition of the leather or oilcloth as it ages, dries out, and becomes brittle. It is equally due to the rotting of the thread that holds the seams together.

When you handle this type of doll, it's best to hold it by the head and the body rather than grasping it at the chest, as would be the natural thing to do. Try to support all of the weight as you move it.

Lay the body out on a padded surface, as you would any other doll. Now any movement of the doll can be accomplished by picking up the ends of the towel and carrying it as in a sling, or hammock, of sorts. This method of transport supports all of the doll's parts equally and doesn't put undue pressure on any one part.

You may wish to use leather cleaner for the body. My own favorite cleaner for these leather and oilcloth bodies (and I have tried many) is the Johnson's Jubilee.

Apply the cleaner amply with your hand or a large cottonball, to small sections at a time, doing one side of the doll.

Allow it time to soak in. If it disappears, apply again.

Pat, rather than rub, the soil off. Rubbing can cause the outer "skin" of the leather to rub away.

Do a second application if needed.

Turn the doll over, and do the other side in the same fashion.

Bisque parts such as the head and arms are done after the leather cleaning, as some of the cleaner will undoubtedly get onto the bisque. Bisque cleaning is explained earlier in this chapter.

Do final cleanup with a cotton swab.

Safe and Gentle Display of Kid-Bodied Dolls

If the doll body seems very fragile, the use of a doll stand is not advised. Some of these dolls will be jointed at the hips and can be displayed in a seated position. If the doll body is very fragile, and is not jointed at the hips, it can perhaps be displayed reclining in a bed, against a large pillow, or in a large chair.

Common Repair Needs of Leather Bodies

Bodies made of these materials seem to suffer the ravages of time and handling more than the ball-jointed strung bodies. Leather ages, dries, and cracks; the oilcloth (sometimes called by the trade name "kid-o-line") is also prone to these same problems. Even when the material itself is in good condition, it is commonly found that the thread holding the seams together has dry rotted, and the body will be ready to come apart at the slightest touch. These dolls must be handled, when necessary, very gently, again underscoring the rule of not handling dolls which we do not own.

The most common repair needs of leather and leather-like doll bodies will be:

1. Reattaching arms and/or legs.

2. Mending tears in the material itself.

Reattaching Arms

There are generally two basic ways that arms are attached to kid-bodied dolls. A third method, less often seen, is also shown. To show these methods more clearly, the drawings show the bodies without the heads in place.

Wired-On Arms

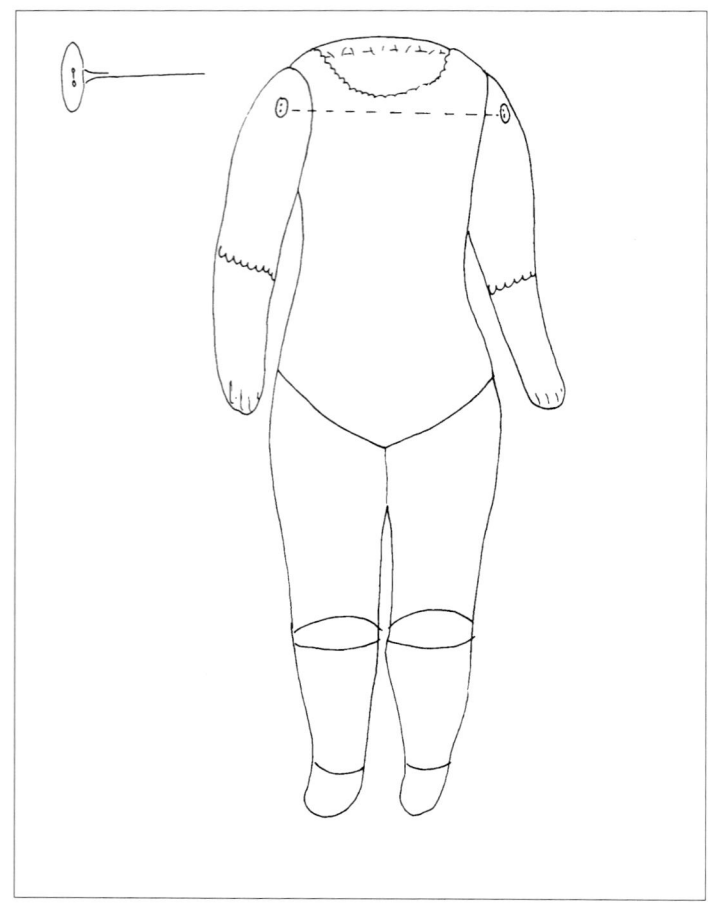

One way that the arms are attached is by being "buttoned" onto a wire which passes through the doll's chest. Metal buttons were originally used for this purpose. Frequently, the button has cut a circular path through the leather or oilcloth and the arm is hanging loose, leaking sawdust, or has fallen off. Another problem can be that the wire is very corroded and breaks, allowing the arm to fall off.

If there is a hole in the leather of the arm, this needs to be mended before re-wiring the arms onto the body. This mend may be done in several ways. The best method is to empty out some of the sawdust from the top of the arm and glue in a small piece of leather or canvas from the inside of the leather arm. If this is not feasible, a small circle of leather or other heavy material in off-white may be glued over the torn area from the outside. Allow this glue to dry overnight, or for several hours, before proceeding.

Measure across the doll's torso and add about four inches. Cut a piece of 22-gauge steel wire in this length, pass it through the metal button (or a replacement plastic button) and crimp.

Then pass this wire through the doll arm, then through the torso (take care that the arm is being placed on the correct side of the body).

Add the second arm and place the button on the wire.

With one hand, compress the arms/torso slightly, trim the wire as needed, and turn it into the remaining hole of the button (compressing the arms and torso allows you to turn the wire back into the button more tightly, for a snug fit of the arms against the body).

This repair may be done with the head attached; of course, take care to work on a padded surface so that the bisque face is not scratched or scuffed by being moved about the work surface. Another safety tip: anytime you are working on a body with the head attached, move any tools that you are not using away from the immediate area. It's easy to slip slightly and bump the head into something which might cause damage.

Sewn-On Arms

Another way that arms are attached to leather and kidoline bodies is by being sewn onto the body lining. All kid bodies have or had, muslin linings inside the torso. These linings allowed for the body cavity to be filled with sawdust or ground cork and sewn shut. The head was then placed over the "inner" body and the outer layer, the leather torso, was then pulled up over the shoulders and glued down.

If the head is off the body, you can see whether or not the lining is intact. If it is dry-rotted away, it is impossible to replace it without emptying out the body and installing a new lining. While this is not an impossible task, it is messy and time-consuming, and will require that you are dedicated to the task It would be accomplished in the same way that the lining is prepared in "Replacing a Torso" in this chapter.

Attaching Sewn-On Arms

Using heavy thread, place running stitches across the top of the leather arm, gather, and wrap around twice with the remaining thread.

Stitch through this gathered/stitched area and "knot off" (pass needle and thread through the last loop of stitching, twice, pull up tightly, and clip).

Position the arm at the shoulder location on the correct side of the body. Adjust up or down so that the hand falls appropriately near the hip. Some kid-bodied dolls have proportionately very short arms. If this is the case, place the arm as low as possible.

Taking larger stitches than normal, zig-zag stitch your way across the lining seamline toward the other shoulder. Rest the needle in the body.

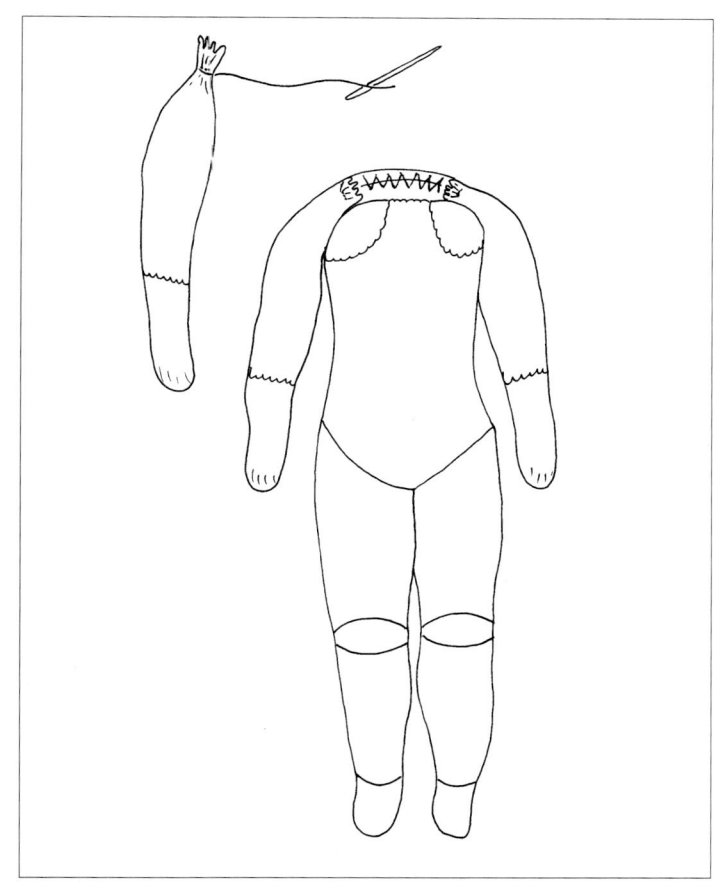

Position the remaining arm at the shoulder location. Take care that it hangs evenly with the first arm. Rubberbands may be used around the arm and body, holding the arms in place while you finish.

Reversing the previous arm attachment, place running stitches across the top of the leather arm, pull tight, wrap twice, and knot off.

Now the head may be glued to the lining, and the leather outer body pulled up, and lightly glued about the shoulders. Use two rubber bands, through the crotch and up over the shoulders, one on each side, to hold the head in place while the glue dries.

Our patient for this arm repair section also needs her legs attached, which we will cover in a later section.

The arms had been sewn to the body outside of the shoulderplate. Since the head needs to be attached, it's a perfect opportunity to re-attach the arms properly (see photo below).

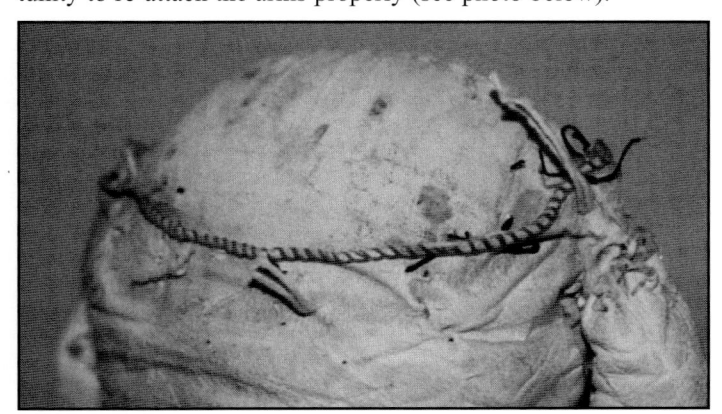

Using a small scissors, the old stitching is removed and the arms are positioned for re-attachment, taking care that their lengths are equal and appropriate to the body.

Using the techniques in the above section, the arms are sewn to the body.

In the following photos you can see how nice she looks now. *Collection of Susan Webb*.

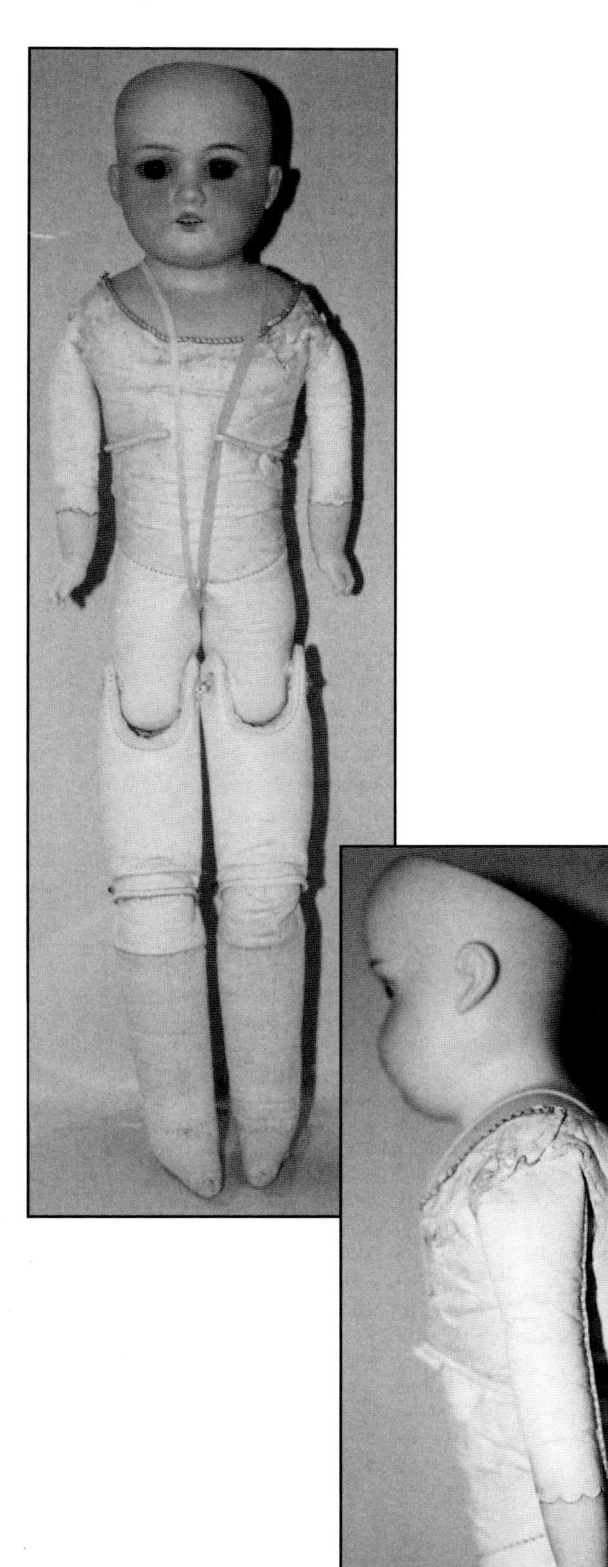

Attaching Sewn-On Arms, with Head on Body

If the arms are off, but the doll's head is firmly glued to the body, the sewn-on type of arm may be attached in an easier "cheater" method. (Hush now, we won't use shortcuts very often!) While this is something of a shortcut, it does no harm to the doll, and is reasonably long-lasting, provided the arms are not moved about often and vigorously.

Using a blunt tool or finger, gently prod against the body under the shoulder plate at the shoulder line. Your finger is ideal for this task, because you can feel if you are beginning to tear the body lining, which may be fragile. The goal is to create a little open space in which to position the tip of the leather arm.

Place a small amount of Tacky glue into this area, making sure it adheres to the bisque. Having the doll upside down will make it easier to see and do this process.

Tuck the leather tip of the arm into this space, using a tool such as a chopstick, if you wish, and wipe away any excess glue with a dampened cloth.

Repeat the procedure for the opposite arm, matching the drop of the fingertips.

Use large rubber bands to hold the arms in place while the glue dries. Allow drying time of several hours to overnight, depending on the size and weight of the arms.

Arms Which String Through Leather Body

Less often, you will encounter a leather body type in which there is a drilled out dowel-type of wood rod through the doll's chest, through which stringing cord is passed to attach the arms. The arms will look similar to other arms on these types of dolls, except that the upper arm is wood with a leather covering. The stringing cord passes through this wooden upper arm and is knotted or held by a pin or nail above the location where the lower bisque arm attaches. These arms are fairly simple to re-string, except that the lower bisque arms must be removed to gain access and install the stringing cord.

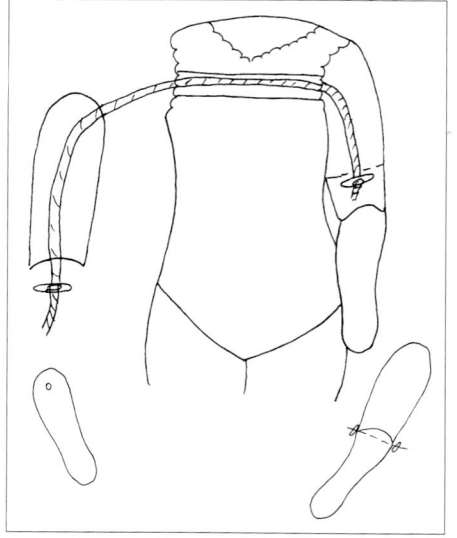

Choose a metal washer in a size that is not larger than the dimension of the upper arm and will fit up into the leather "sleeve".

Place a knot in one end of the stringing cord. String on the washer and pass the cord up through the upper arm and out the top. The washer will prevent the knot from slipping through the opening in the arm.

Pass the cord through the wooden "tube" in the doll's chest, then through the other arm in reverse order. Pull the stringing cord as tight as you can manage and crimp off with a hemostat. Place a metal washer on the cord and then knot the cord up to the washer. Pull the knot tight, trim the stringing cord end, and then release the hemostat. Now attach the lower bisque arms.

Attaching Legs which are Wired Through the Body

The photo below shows a sweet little doll whose legs need to be reattached. Poor thing, not that she need worry for long. This is a simple and straight-forward task.

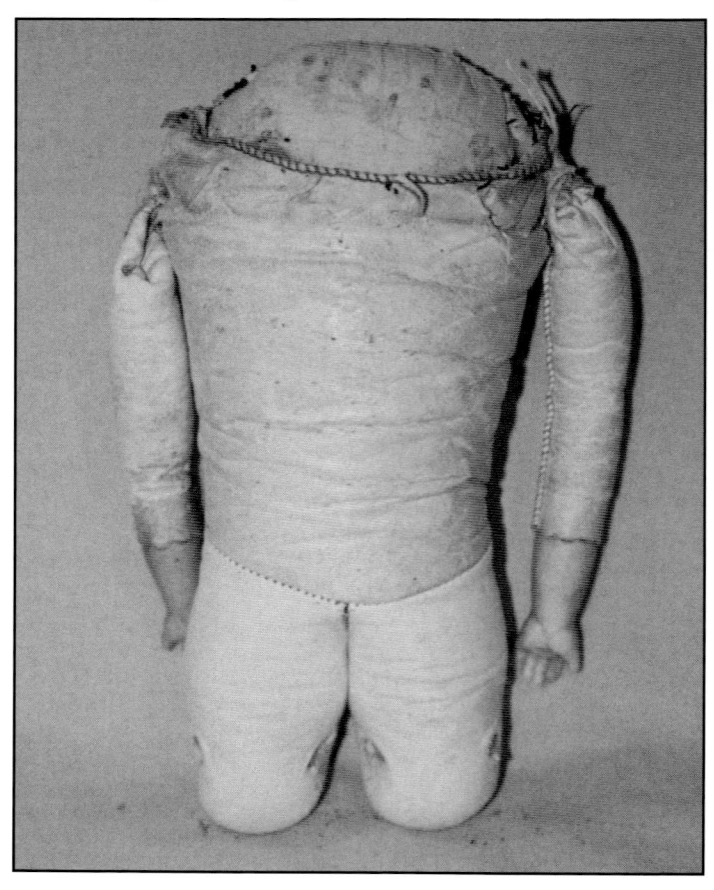

A clear button is placed onto a length of 22 gauge steel wire and crimped.

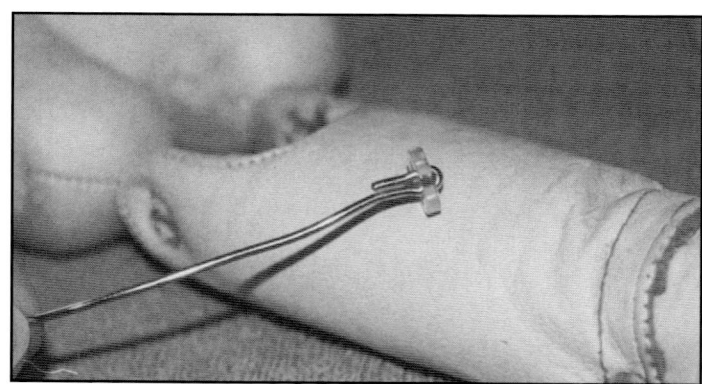

Working from the center out (it's easier to crimp the remaining button if it is more accessible on the outer, rather than inner thigh), pass the wire through the inside lower leg, then upper hip, then outer lower leg, as shown.

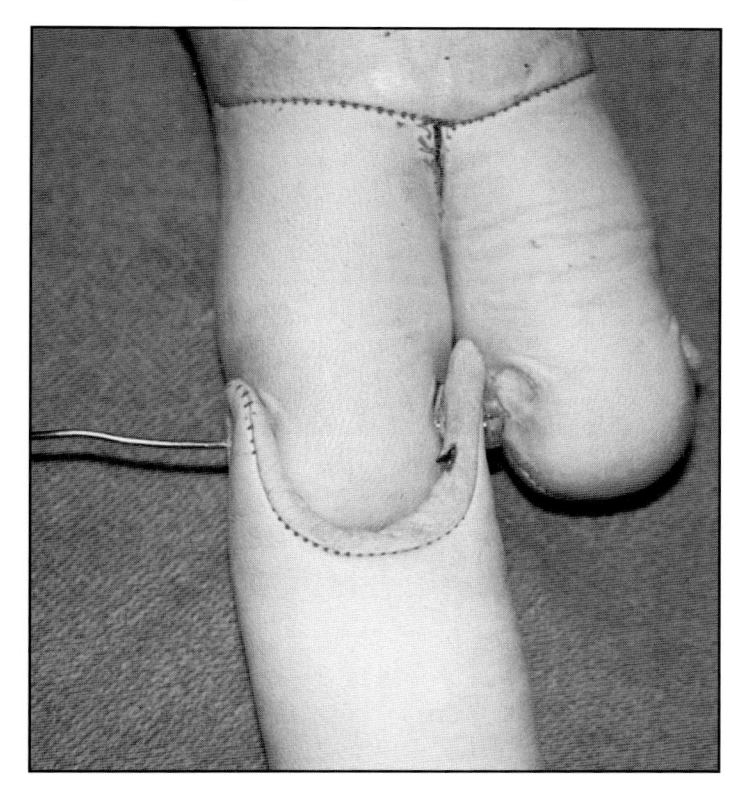

Cut and crimp the wire onto the outside button. Repeat the procedure for the other leg.

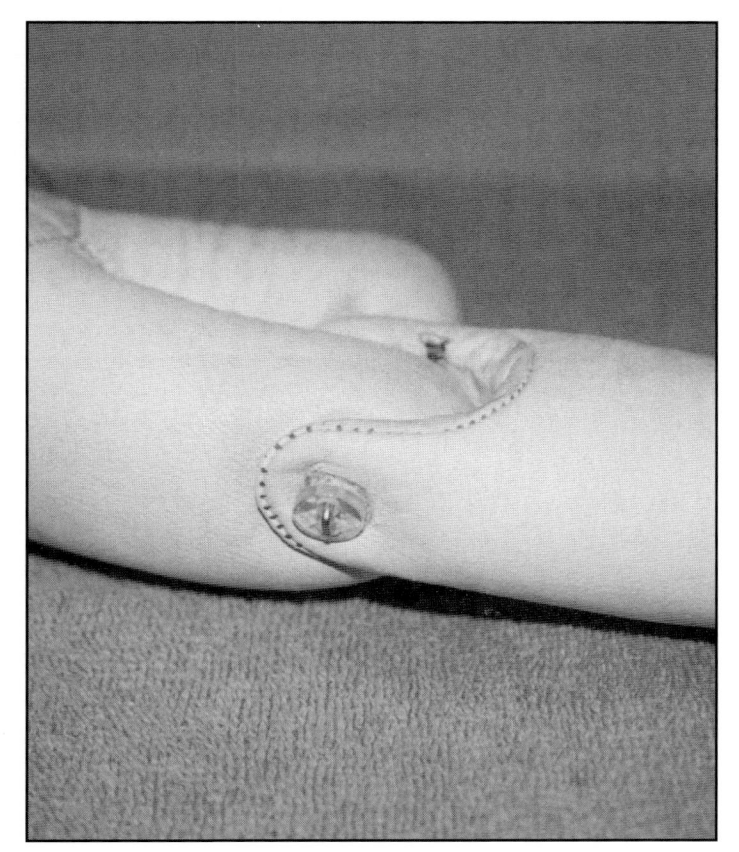

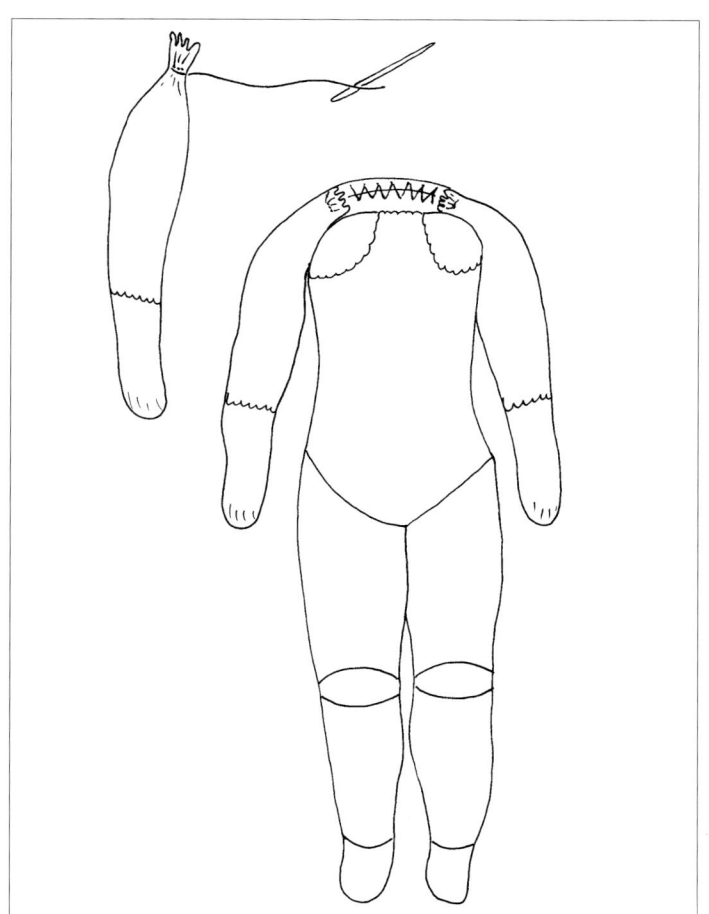

A full-length sketch of the finished doll is shown above.

Note: this leg stringing is often seen done with one wire rather than two, buttoned only on the outside of each thigh. While this is workable (and less work!), it leaves the inner thigh holes (where the wire passes) without anything to prevent wear and the leakage of stuffing material. Hence, the additional two buttons are worth the effort.

Mending Tears/Seams in Leather

In the following photographs, the sequence of body repairs begins with cleaning and conditioning a kid leather body which is in quite poor condition.

The body is laid out on a plastic sheet to contain the spilling sawdust. Using a little polyester or cotton stuffing at each opening helps to contain the sawdust and gives a little extra firmness against which to sew.

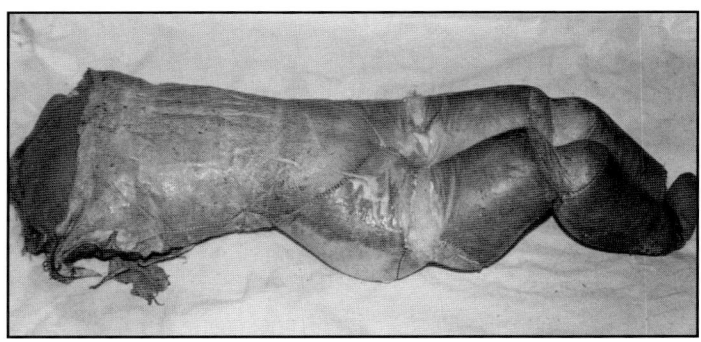

55

The top side of the body is cleaned gently, using ample amounts of Johnson's Jubilee or a leather cleaner of your choice. It looks like it will be best to clean and then mend this side before turning the body over, as the holes are quite large and much more sawdust would spill out! The cleaner may be applied with a cottonball, left to soak in for a few minutes, and blotted away. Care must be taken not to scrub away the "skin" of the leather by using a brush or rubbing too vigorously.

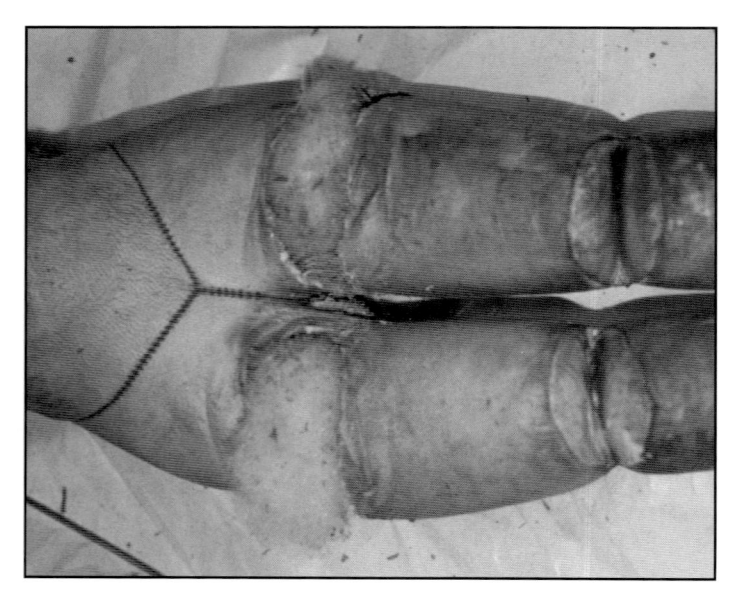

A rolled up towel helps to hold the leg in position so that the leather is not strained. Begin sewing the "wound", using a double strand of thread, hiding the knot inside the opening. The type of stitch used will vary with the type of opening, the access available, and whether you are right- or left-handed. The stitch used here is a ladder stitch, taken rather deeply, as the leather is quite frail and shallower stitches would pull through the delicate leather.

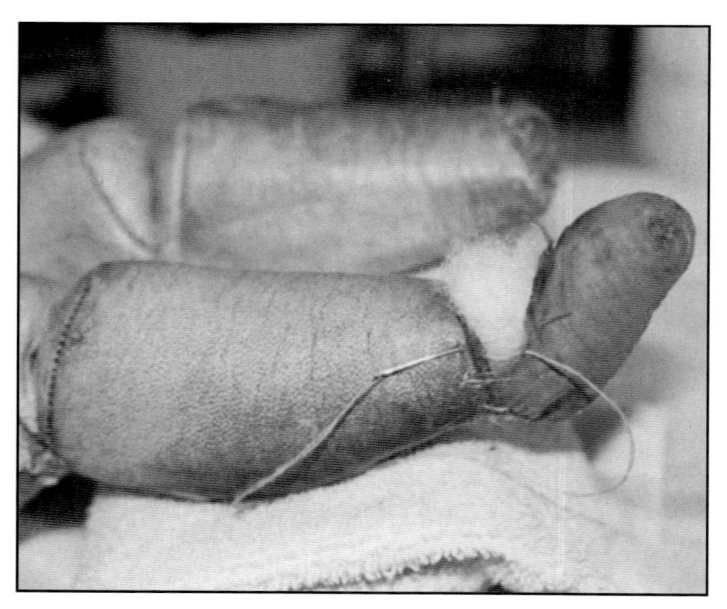

The foot is held in position as the sewing continues around the torn seam.

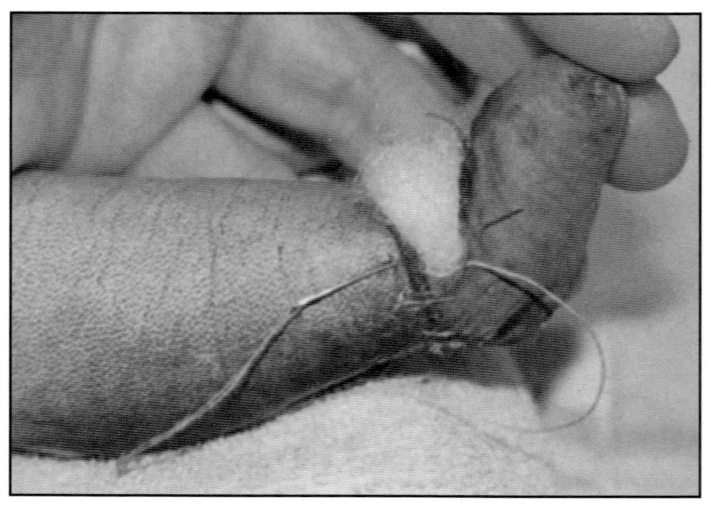

For this tear in the hip area, it was easier to begin sewing on the inner side of the leg and proceed to the outside.

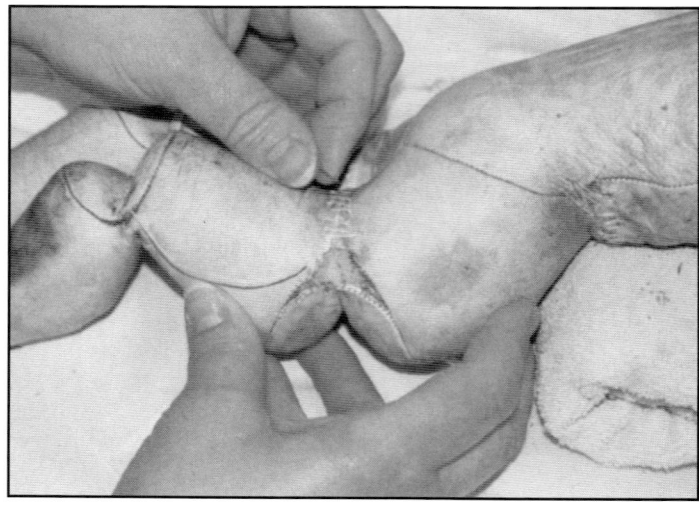

The same type of ladder stitch is being used heading into the gusset area, which will require bringing together three "corners" of the opening.

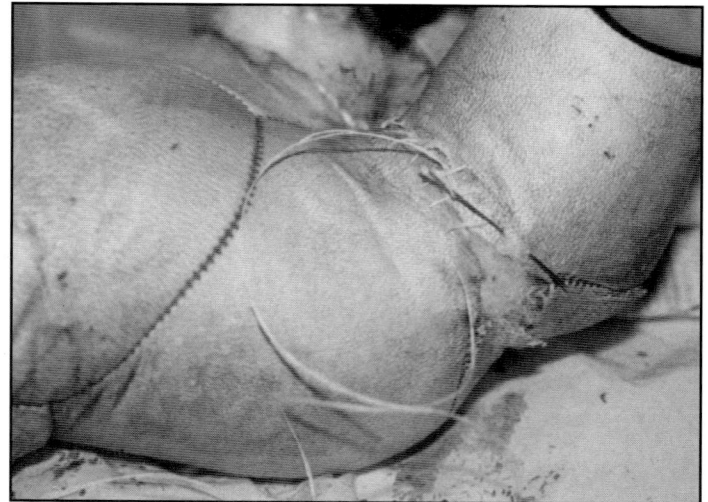

The corner area is closing up nicely. Now the body can be turned over for the cleaning of the bottom side.

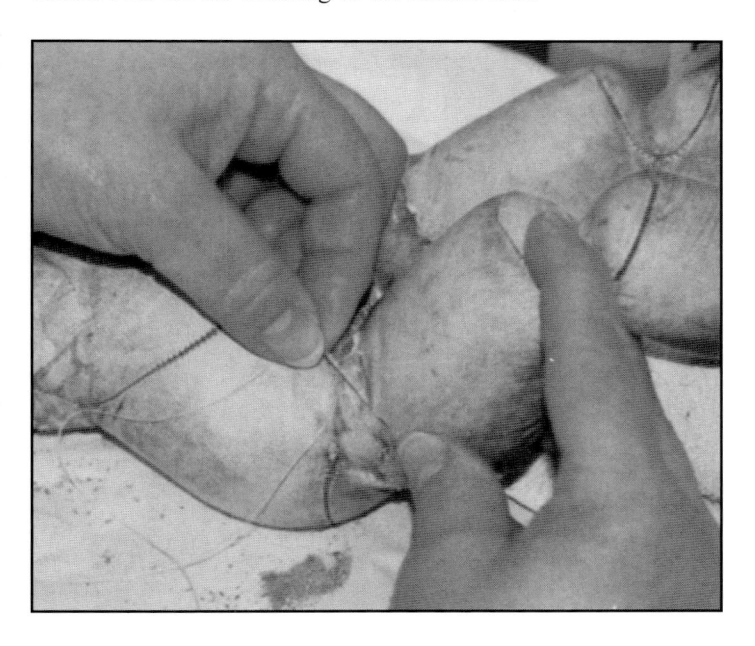

"Upholstering" a Damaged Leather Body

Antique Kestner #147, deteriorating cloth on upper torso; the leg on the right shows the benefit of a proper cleaning.

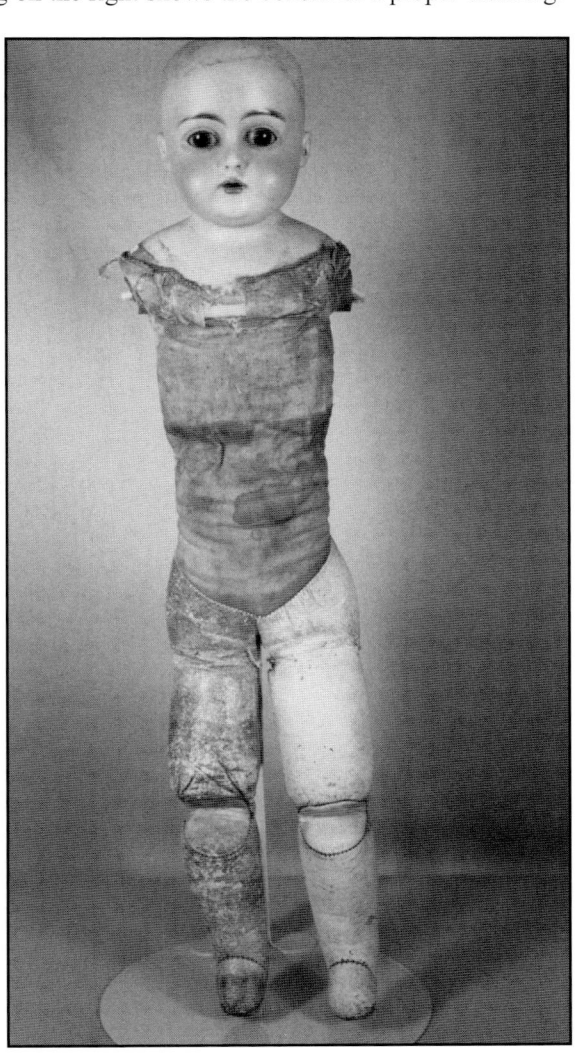

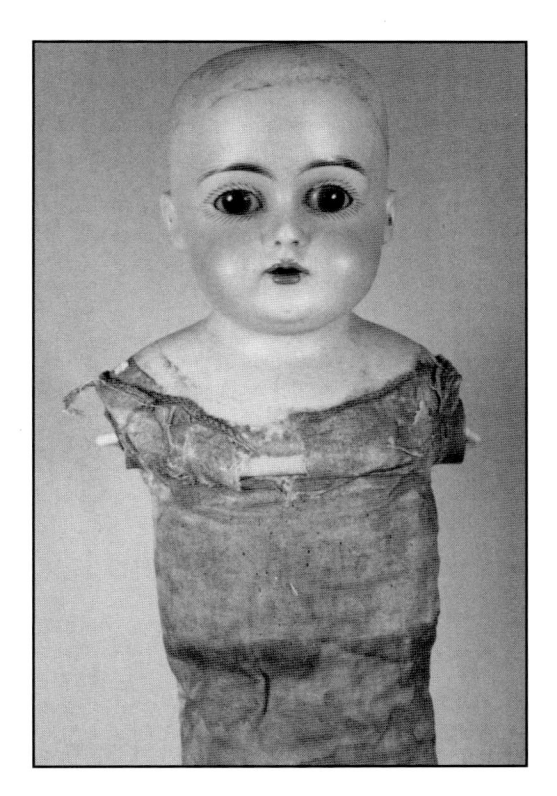

Close-up reveals a very pretty face under the soil and glue and a fragile body.

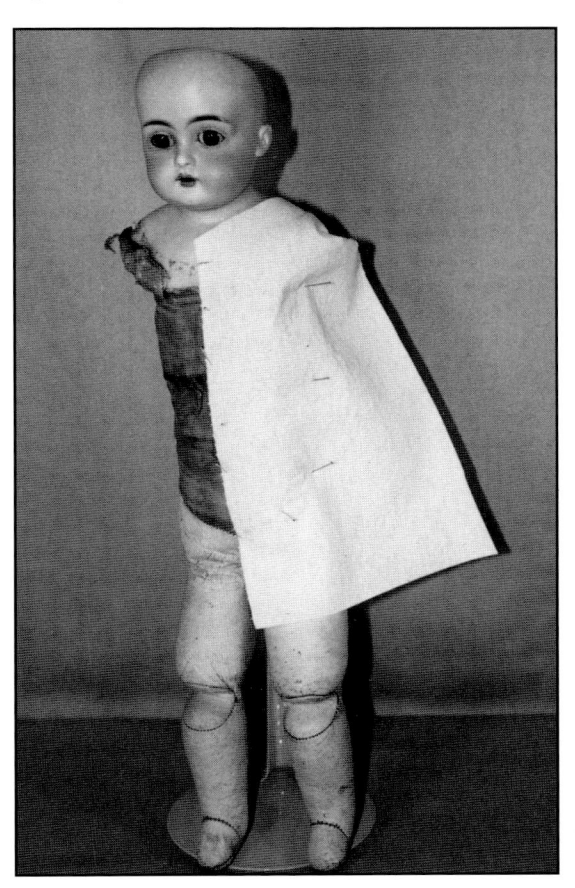

The remaining leather and the head have been cleaned; paper toweling is pinned at seam lines to make a pattern to replace the damaged original cloth portion of the body.

57

By "connecting the dots" of the pinholes in the paper towel, an exact pattern is obtained for the front and back body pieces. These are cut from a broadcloth fabric which has been heat bonded to give it the approximate sizing of the leather. The front and back pieces are seamed at the lower edge and joined at the side seams. The arm and neck openings have been cut with scalloping scissors to duplicate the look of the original body. The bonding allows this cutting to be done with a smooth edge. Stay-stitching 1/4-inch from the edge of these openings gives it a finished look.

Left: After assembly, torso covering is lightly stitched over the original cloth portion.

Center: Original arms are re-wired onto the body.

Right: With an appropriate (new) mohair wig, she is beginning to look like the little treasure that she is!

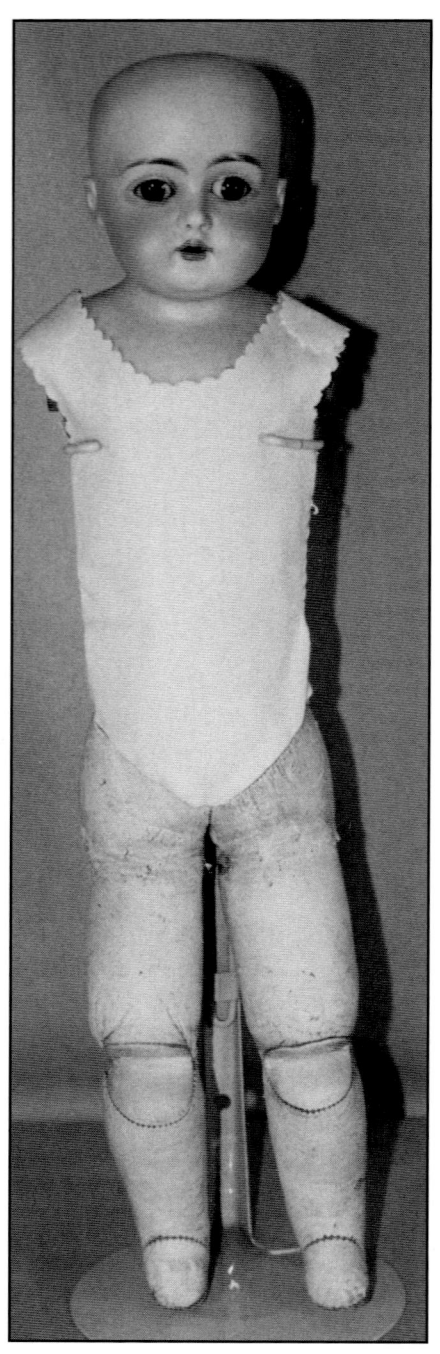

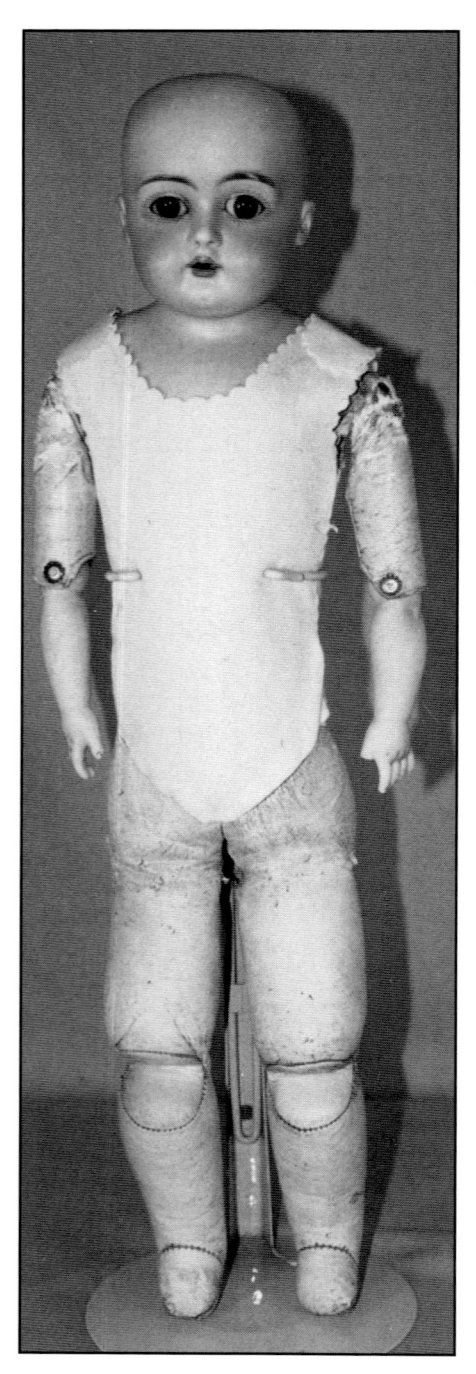

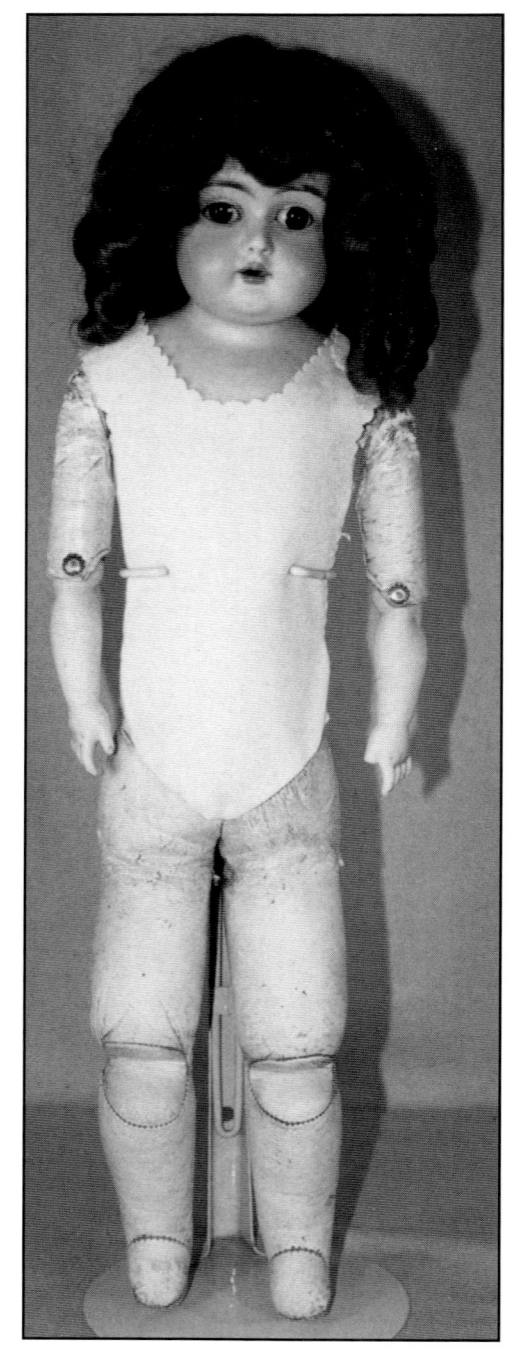

Replacing Badly Damaged Cloth Torso

A much-loved little doll is brought to us with the sole desire that she be brought back to life in whatever way is possible. She is indeed another little "basket case", a small unmarked German shoulderplate head on a cloth and leather body.

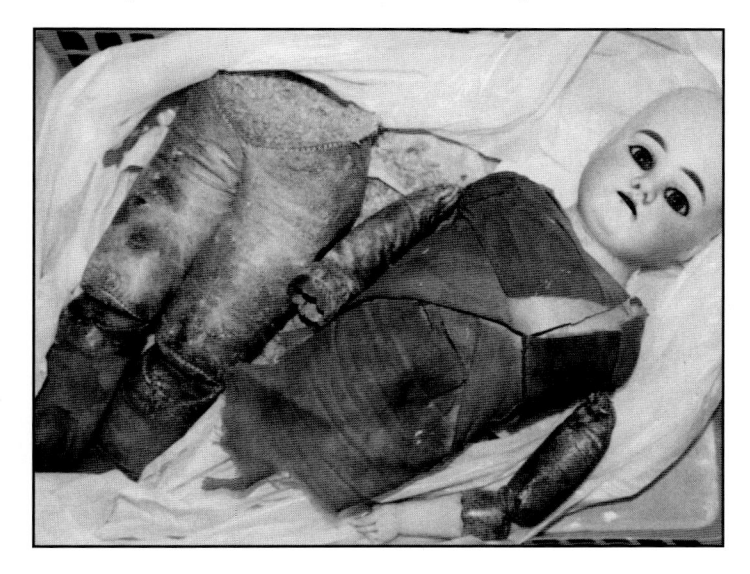

This doll's cloth upper torso is basically dry-rotted away, the leather arms and lower body are quite soiled.

The plan is to clean and condition the leather portions remaining, clean the bisque head and arms, make a copy of the upper cloth body, and reassemble the doll.

The lower leather portion of the body is cleaned gently with Johnson's Jubilee cream, allowing time for the cleaner to penetrate before swabbing off with cotton balls. The dark soil being removed is undoubtedly due to the coal heating which existed in homes in earlier years. This type of soil usually takes several applications of cleaner (see below, top left, right and bottom photos).

A noticeably cleaner body is now checked for needed repairs. A small tear in the fabric lower leg is mended with (almost invisible) nylon netting and Tacky glue. The glue is brushed onto the torn area, an oval-cut piece of netting placed into the glue, and then the glue brushed out through the netting, making sure it adheres all around. Drying time for this type of mend is usually no more than two or three hours (see bottom right photo).

Copying a Damaged Cloth Torso

It's a little tricky to reproduce a part that is so rotted away. We approach this problem by judging the doll's approximate height, which we decide is about 16 inches. This allows us to draft an approximate torso pattern from a similar doll in our collection.

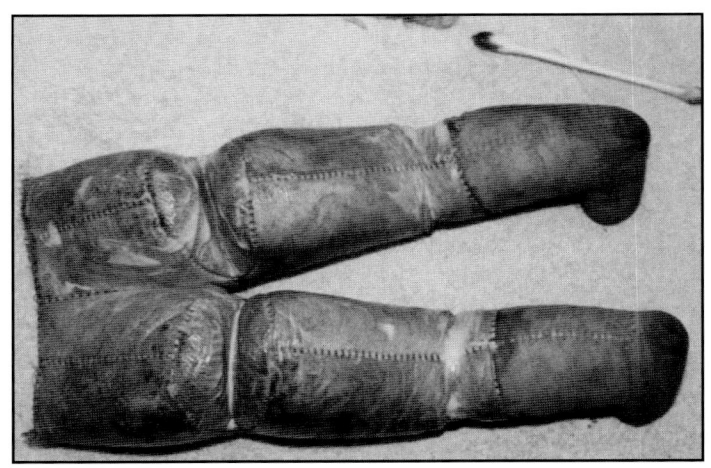

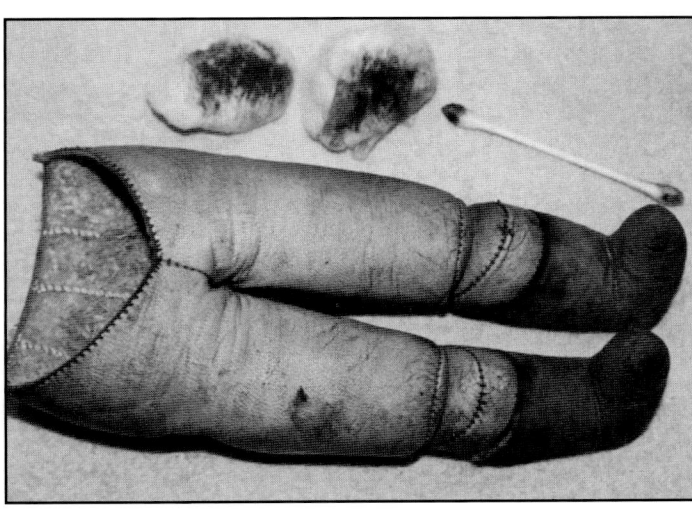

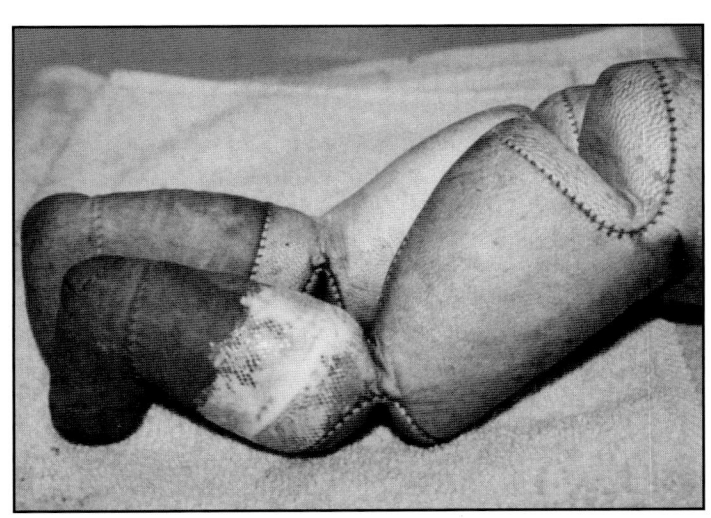

We take a back waist measurement (see below). This photo was taken before the part was cleaned! Luck is with us, the measurement is exactly the same as that of the comparison doll we are using for the body pattern.

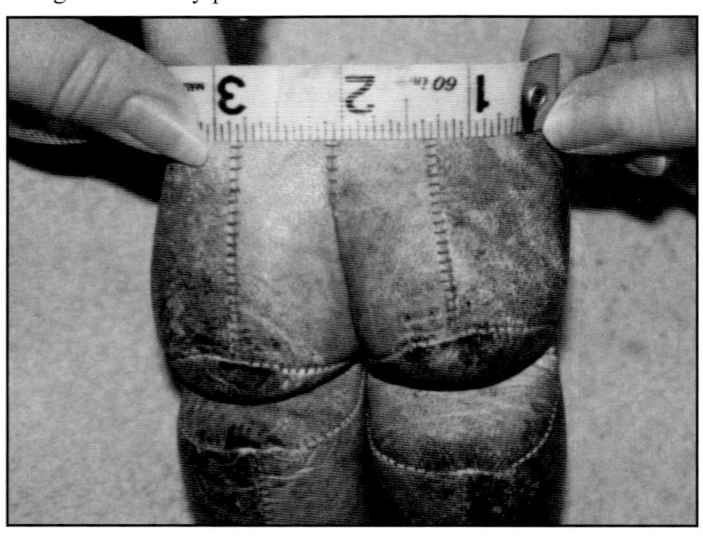

The patterns for the body torso lining is shown below, top sketch. The body torso pattern is on the bottom.

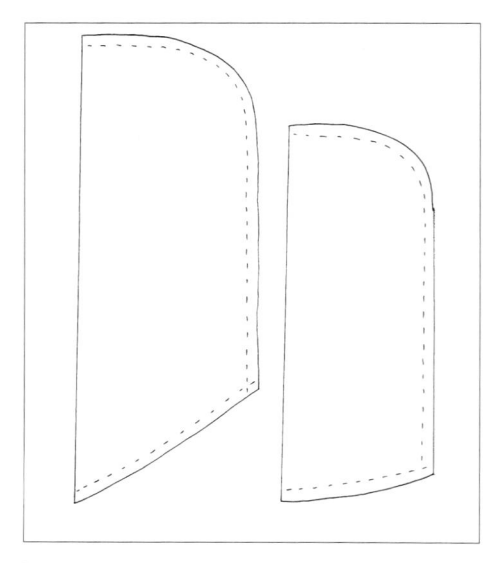

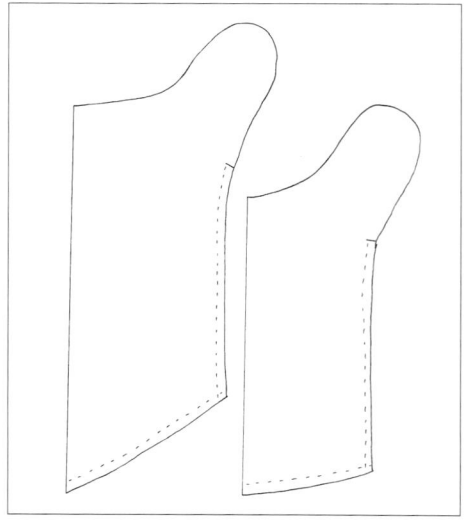

Preparing Fabric for Torso:

We will use an iron-on bonding material, ironing it onto a white fabric similar to cotton duck. We chose this fabric because we felt it would have the most similar look to the original body, with a bit more weight, similar to the weight and feel of leather, once bonded.

Iron-on bonding is a method used to create iron-on appliqúes. It makes it possible to cut a shape from the fabric without fraying edges. While the usual use of this involves peeling the backing paper and ironing down the appliqúe to its background, it is also helpful without using this last step. The benefit of using the iron-on bonding material for our project is that the upper edges can be scallop-cut without having edges that will fray. We will simply remove the backing paper and skip the last ironing step.

The body lining is cut from white broadcloth.

Assembly is as follows:

Sew the torso front lining to the outer front torso, at the lower edges only, right sides together (see top left sketch).

Sew the back lining to the outer back torso, turn right side out and press (see top right sketch).

Right sides together, sew the side seams, body and lining up to the underarm area, as marked on the pattern (see bottom sketch).

Turn the finished torso/lining combination right side out and press the seams.

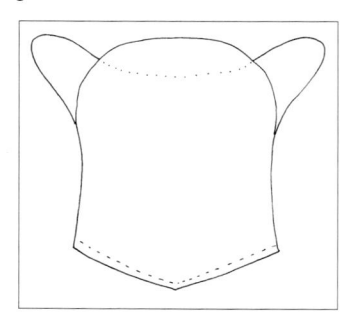

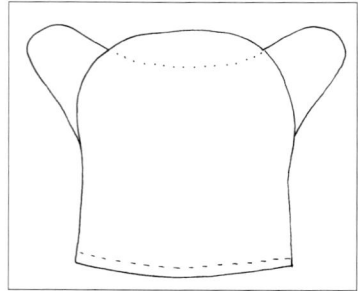

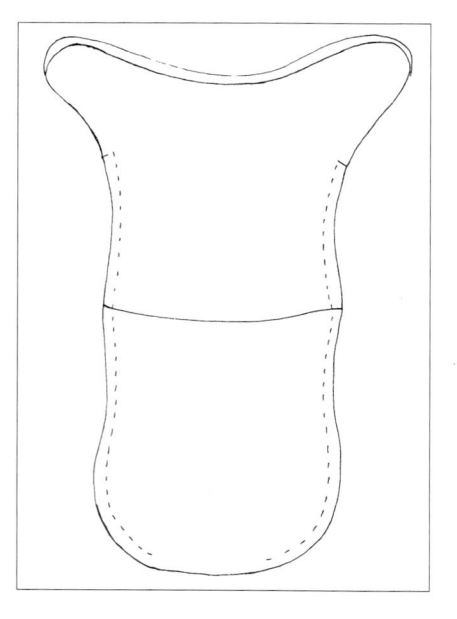

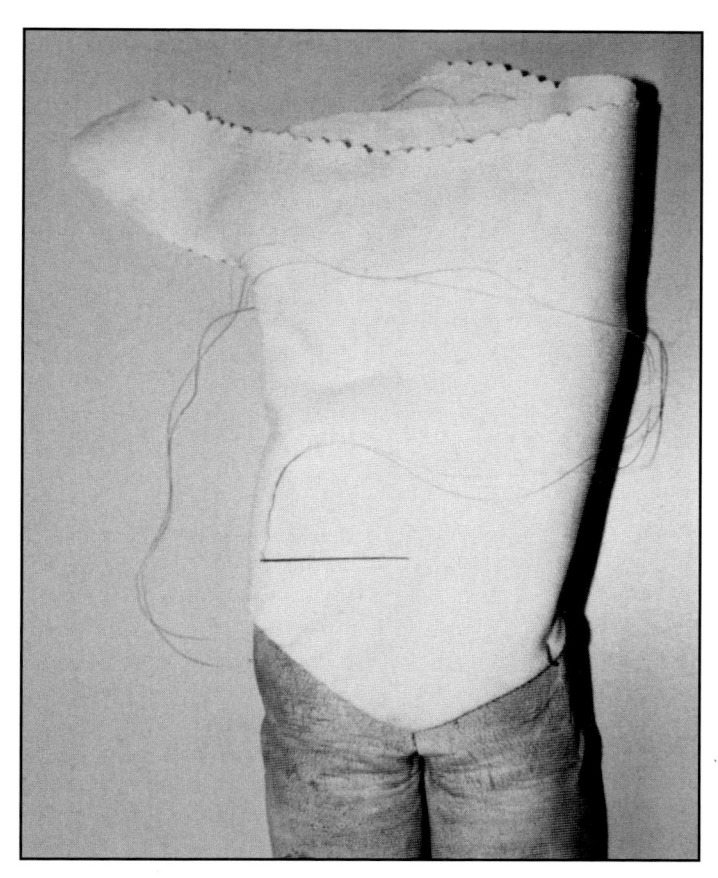

Hand-stitch the finished lower edge of the upper torso to the lower torso, using heavy thread.

Using a wide-mouth funnel, begin adding the old sawdust into the body cavity.

Use a tamping tool (easy to make one with two sizes of dowel rod) to pack down the sawdust tightly.

Fold down the outer torso "tabs" to give better access to the edges of the lining. Bring the edges of the lining together and pin to hold securely while closing the opening, using heavy thread.

Stitch through the top of the upper arm, pull to gather slightly, and stitch again through the gathers (this method is shown under "Sewn-on Arms", in this chapter).

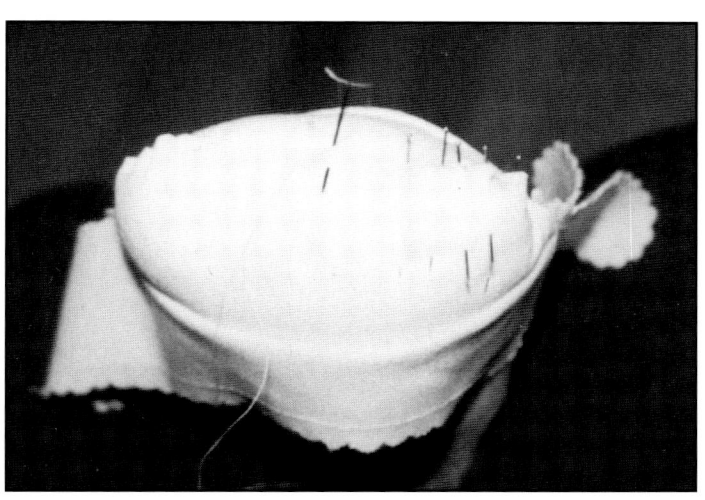

Attach the arm to the upper torso, positioning it properly in the armhole opening. Take slightly larger stitches across the lining seam, working your way across to the opposite shoulder area.

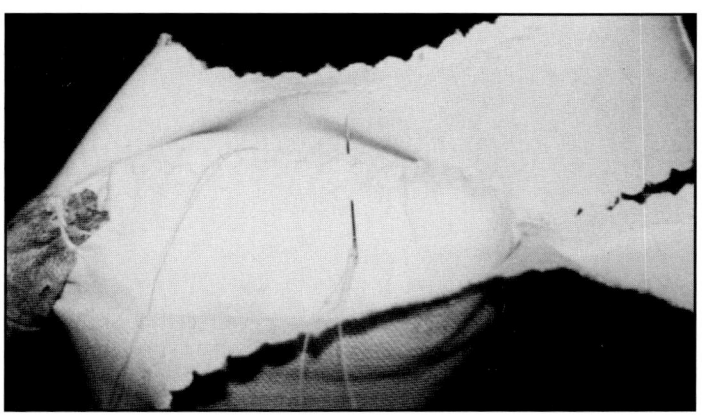

61

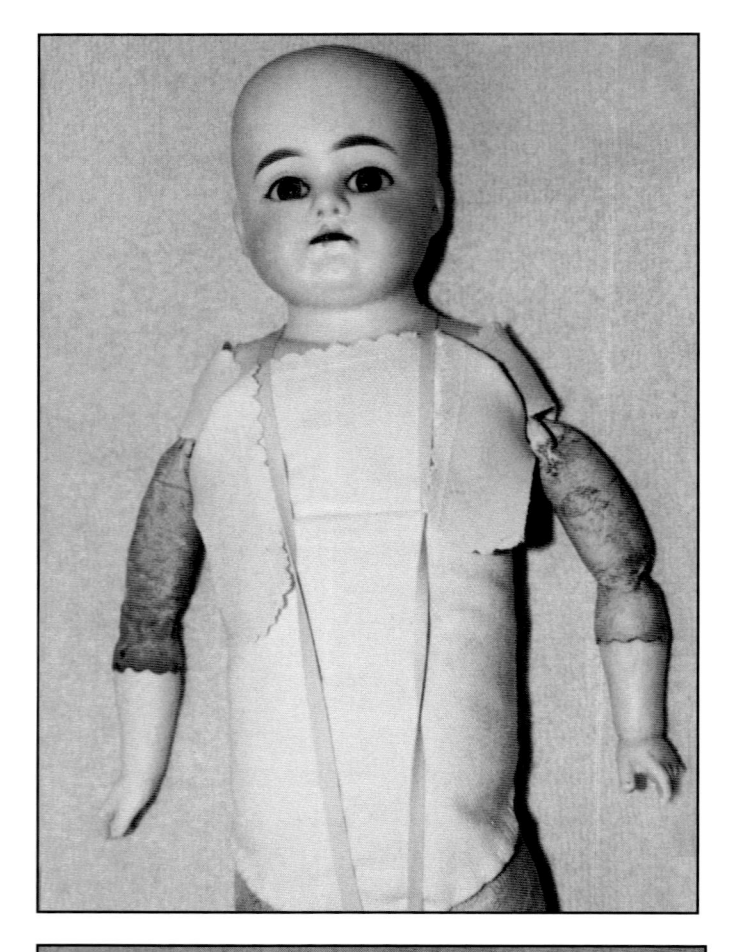

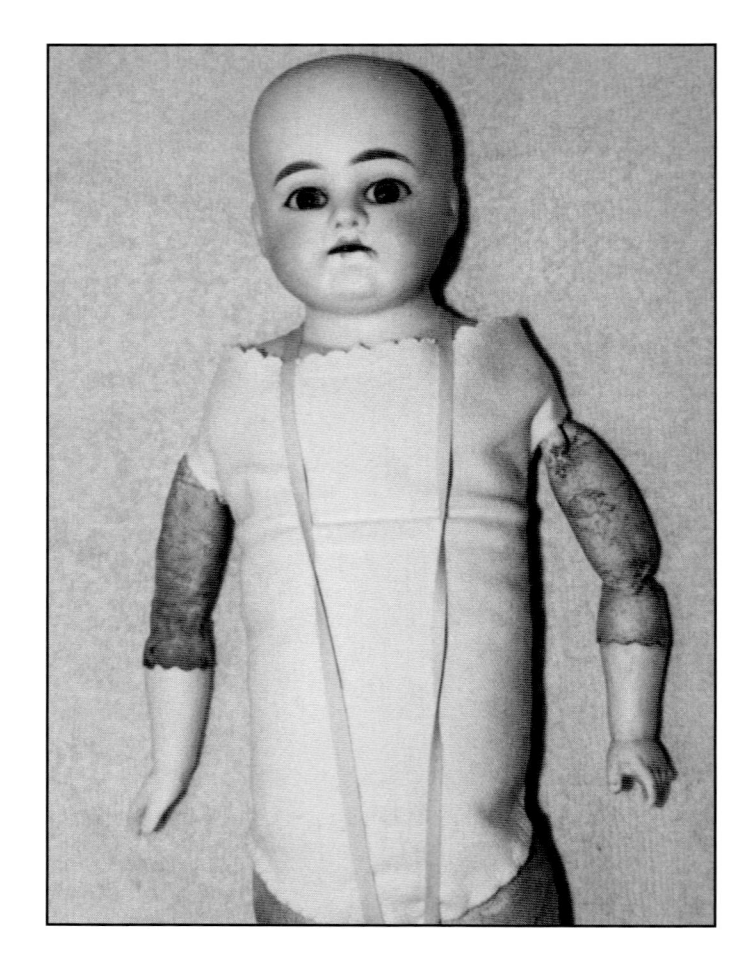

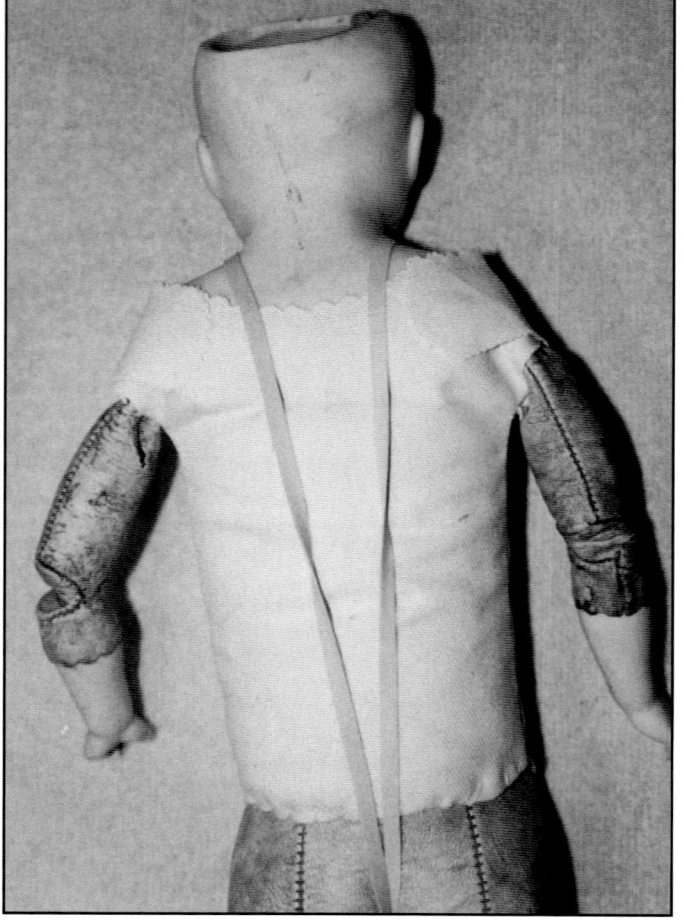

Attach the second arm in the same way, reversing the steps. Take care to position the second arm so that the drop of the hands is even.

Leave the outer torso "tabs" folded down. Place two large rubber bands in the crotch, one coming up on each side of the body.

Position the head on the top of the torso between the lining and the outer torso. Check the tabs for excess length; they may be trimmed with the scalloping scissors if they are too long. When you are satisfied with the fit of the tabs about the shoulderplate, proceed as follows:

Glue the shoulderplate in place with Tacky glue and slip the rubber bands over each side of the shoulderplate to hold the head in place while the glue dries (see top left photo).

Lightly glue, with Tacky, the outer torso tabs into place; first the back tabs, forward, and then the front tabs, toward the back, over the back tabs (see top right and bottom photos).

Now this little German bisque has a body, and she's ready for a wig and some cute clothes! Collection of Janice M. Kast.

Replacement Bodies for Shoulderplate Bisque Dolls

Sometimes a doll comes along badly needing a replacement body, and her size is such that finding a correct replacement antique body is highly unlikely. Such is the case with the antique

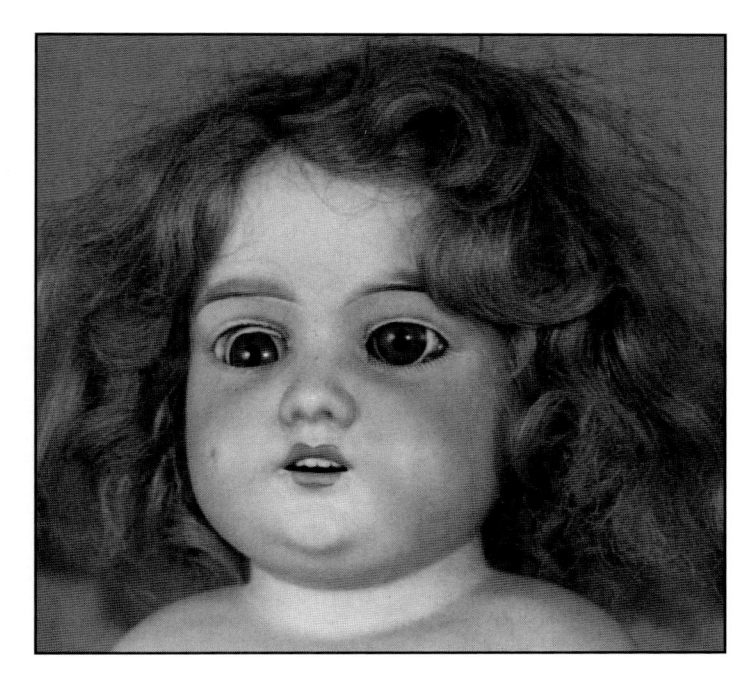

Armand Marseille #370 in the START photographs on this page. The head of this doll measures about 16 inches in circumference and is over seven inches tall from crown to lower edge of the breastplate. With original glass sleep eyes, mohair wig, multi-stroke brows, lovely blushed cheeks, and no breakage in the bisque, she is certainly worth restoring. Sometime in her past, the head was put on an American made 1930s cloth baby body with composition hands!

Of course this doll represents a girl circa 1900. Her original body would have been a leather child body. On a properly sized body, she will stand nearly 30 inches tall, and properly coiffed and costumed, would be a focal point of any room.

Since we were not hopeful of acquiring the proper size antique body in a reasonable time frame, we opted for the feasible and economic alternative, a new "kidoline" body. These are available in many sizes.

The body is tightly stuffed, and the torso lining sewn shut with heavy thread (see photos below).

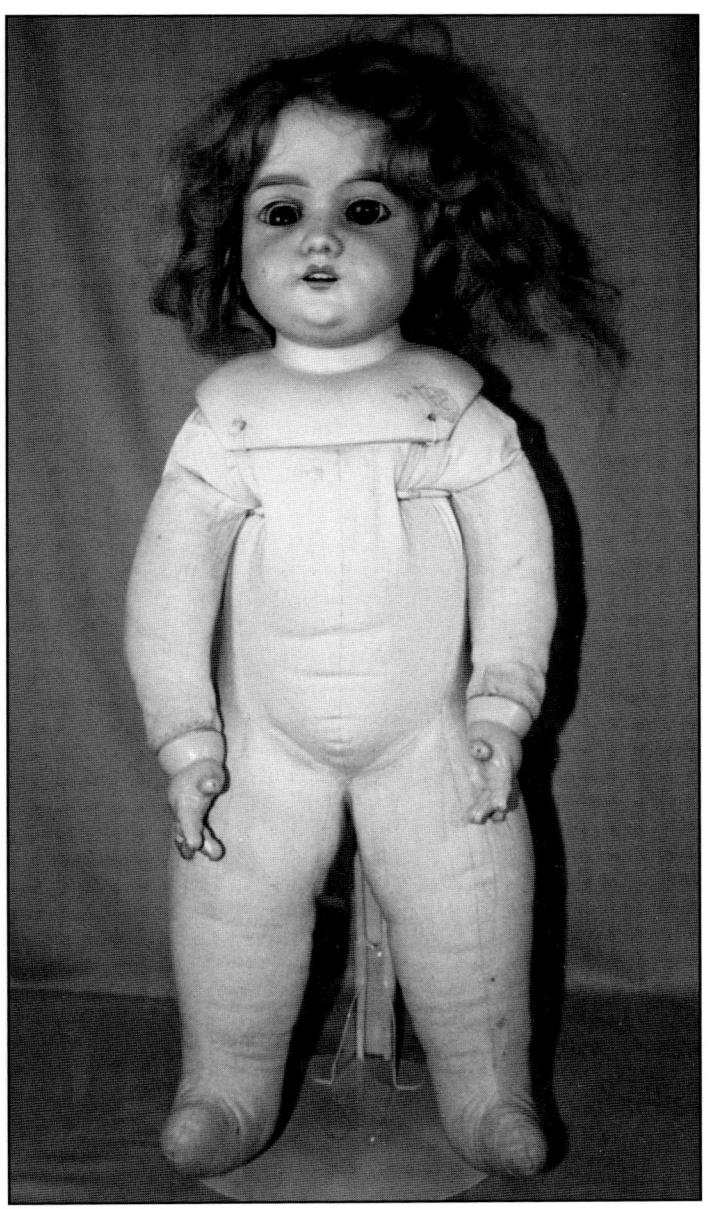

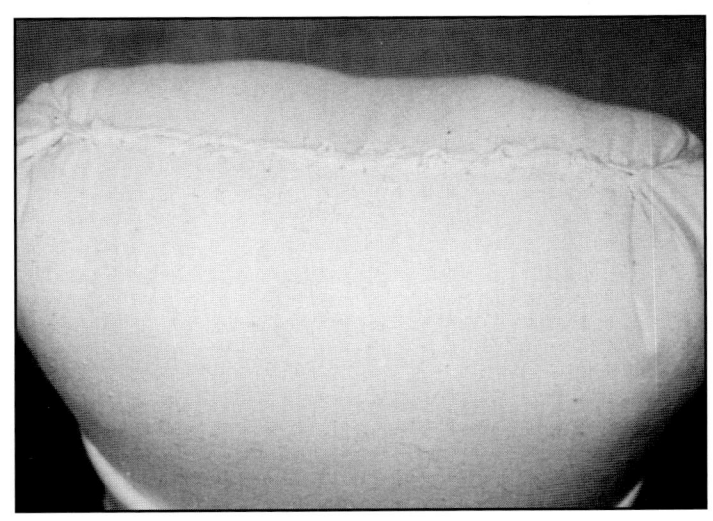

Replacement bisque lower arms are inserted and glued into the kidoline arms

The upper arm section is then stuffed about 2/3 of the way. Stuffing all the way up will cause the arms to stand out from the body too far.

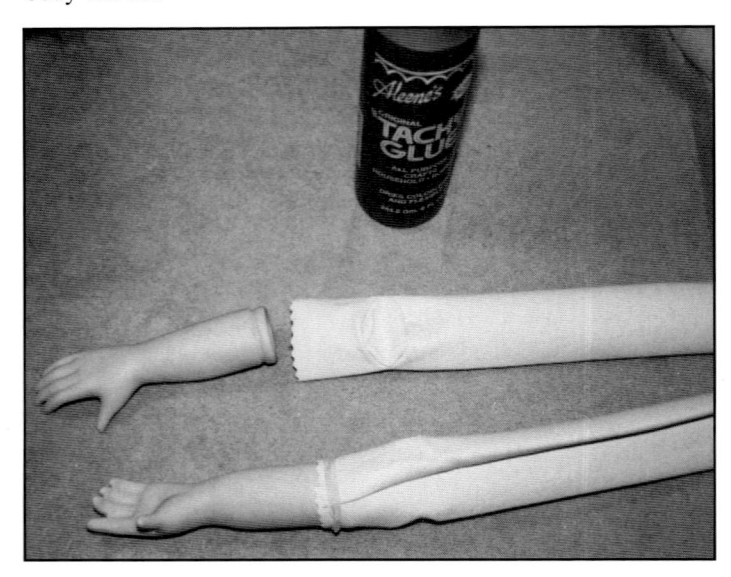

This close-up of the face shows the eyes to be quite out of alignment. This repair is described in Chapter Five. Having finished the eye repair, we complete the assembly of the doll. Extra care is taken in gluing on such a large size head.

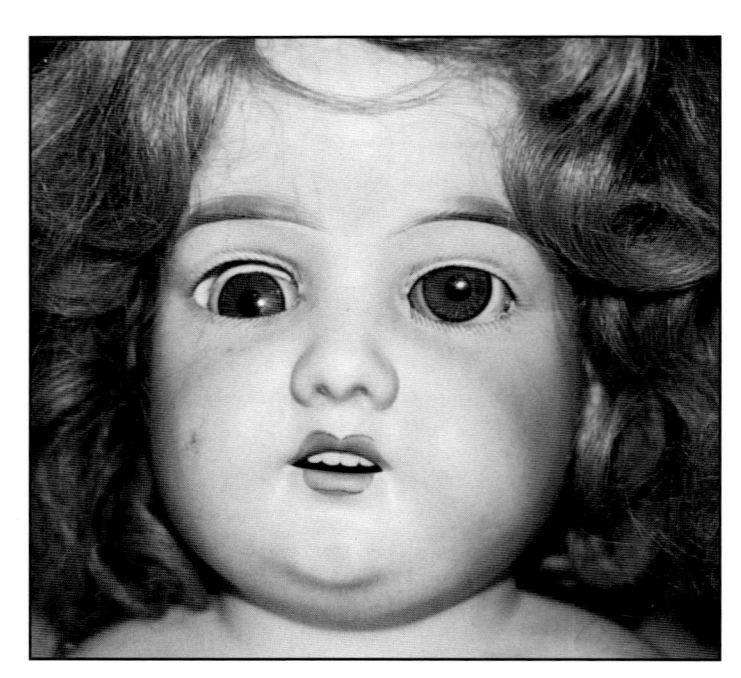

The assembled doll, at right. *Collection of Sara Bruns.* While not restored to original condition, this doll now enjoys the best alternative that we could give her, a body of proper style and size. Since this option is not expensive, it remains feasible to replace this new body with a correct old one, should such a body ever become available. In the meantime, this doll can now be properly displayed and enjoyed by all who see her.

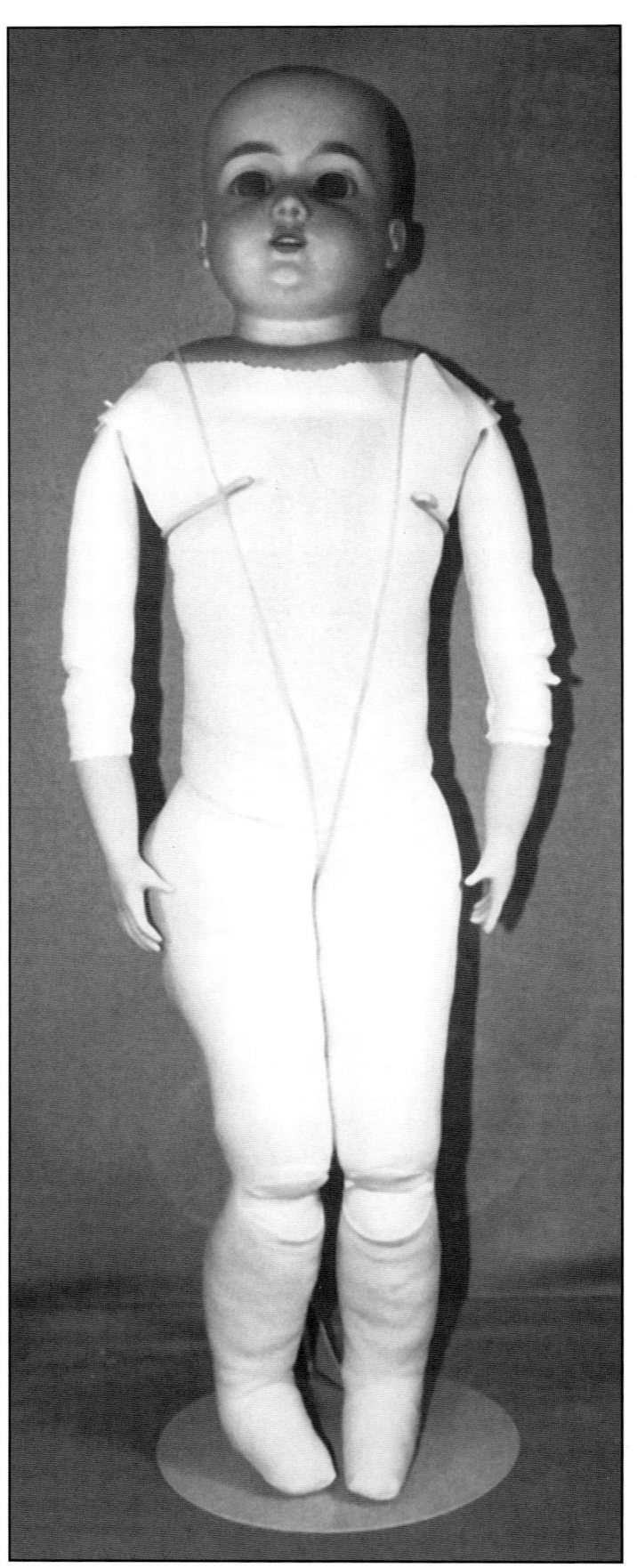

Chapter 5

Eye Repair in Antique Dolls

Of all the types of doll repair, eye repair is perhaps the most challenging and the most critical to the appearance of the doll. "Bad hair" can be tucked up or pinned back and hidden under a pretty bonnet. A damaged, mismatched, leaking, or poorly strung body can be disguised when the doll is dressed in a charming outfit. The eyes of a doll, however, are obviously noticeable from the first moment. In fact, it is often the soulful eyes of a doll that beckon you from across the room.

So the eyes have it....or else!

The successful doll buyer or restorer needs in his or her arsenal enough knowledge to make right decisions and to stay "out of trouble", that is, with regard to buying a doll with which regrets will surface to spoil the fun.

The basics of doll eye knowledge would include:
Knowing what type of eyes the doll should have
Knowing how to identify problem installations of eyes
Knowing what can be easily corrected and what will be more difficult

We touched on types of doll eyes in Chapter Two. To fill out this area of information, we will state some very general categories regarding doll eye types, which will serve you well in most instances.

Types of Antique Doll Eyes

Most of the German dolly face dolls had glass sleep eyes, with or without hair eyelashes.

"Sleep" eyes are mounted to a T-shaped wire with a lead weight. When the doll is laid down to sleep, the eyes close. These eyes are held in place by plaster "cups" at each side, which are notorious for falling out of place when they are dried out with age and subjected to a sudden blow or shock. Often, they break inside the head when this happens. This explains the rule about NEVER shaking an antique doll to open or close the eyes!

Other German dolls, mostly shoulderplate types, had fixed, or "set"" eyes, which did not move. These eyes, either regular round glass or flatter oval glass were positioned into the head with a temporary adhesive material and covered with wet plaster. The plaster adheres to the porcelain head as it dries and holds the head in place.

Some earlier (before 1880) German and many French dolls had paperweight eyes. These are glass eyes which have an additional bulge or "dome" of glass over the iris to magnify it (hence the name paperweight) and give it beauty and depth. Paperweight eyes were usually beautifully "threaded" with tiny white lines in the glass which made for a realistic look. Paperweight eyes are always set eyes. They cannot be made to close, as the dome shape will not clear the eye opening.

Recognizing Eye Problems

The most essential piece of equipment for this type of work is a magnifier glass, the type which can be worn as a pair of glasses, allowing you the free use of both hands. You will be amazed at what you can determine about doll eyes that you couldn't see before, even with good vision.

Breakage

It is not uncommon for doll eyes to be in place and functioning in the head and still be cracked. Often, a crack will exist right at the line between the iris and the white of the eye. It is well concealed by the difference in color. If buying a doll or preparing to work on one, you will want to know if the eyes have hidden hairline cracks.

Improper Installation of Set Eyes

The worst of doll eye repair sins, because it is so unnecessary, and so unattractive for the doll, is the leakage of plaster from the setting of the eyes, visible in the eye rims.

When doll eyes are set fixed, sticky wax must be used to temporarily hold them in place. This wax is readily available in stick form from doll parts suppliers. It is translucent white, pliable, and somewhat sticky to the touch, something like a licorice stick. A "worm" of this wax is placed about the eye and the eye is placed into position. The wax holds it firmly enough, yet allows it to be moved about for the best gaze and position. Wet plaster is then applied to cover the eyes, and as it dries it, adheres to the porcelain and holds the eyes in place. The other major advantage of using the sticky wax is that it prevents the plaster from running through to the front of the eye and settling around the rim, for all the world to see.

Incorrect Type or Size of Eye

The eye, of course, needs to be large enough to fill the eye opening. More importantly, the iris of the eye needs to be large enough that its color touches both the upper and lower rim of the eye opening. If the iris is too small, the doll will look surprised or scared. If too large, it will simply look "wrong".

Choosing the correct eye type is simple. If the eyes must be replaced, it is better to use new glass eyes than antique eyes which are not the right size. There will also be the discussion of using acrylic (oh horrors!) eyes in an antique doll. There is a considerable price difference between good glass eyes and acrylic eyes, so we will step out of the fray and simply say...if you must economize and use acrylic eyes in an old doll, buy an attractive, well threaded pair. They are inexpensive and will pass inspection better than the flat oval unthreaded plastic eyes that seem to be everywhere.

Since we are discussing eye installation mistakes, we must stress the use of proper materials so that the eyes are removable if the need ever arises. Do not use grout or wood putty to set eyes. It is very difficult, if not impossible, to remove these materials without risking the breakage of the doll head. This is why grout is so great to use when tiling the bathroom...it holds virtually forever. Plaster, on the other hand, can be soaked loose from a porcelain head in a few minutes with warm water.

How can you tell what was used to set the eyes in a doll? Of course, you need to be able to look inside the head, that is, the wig and pate must be removed. Plaster is white to very light gray and even in color; grout is usually white or gray, but coarser and "sandy" looking, almost like cement. Beware. Wood putties would usually be yellow to light brown and more easily identifiable.

Another eye repair failure, one simple to detect and avoid, is a mismatched set of eyes. I recently came across a doll which had two different eyes. They were different shades of brown, different sizes, and one was paperweight while the other was a round eye. Poor Dolly. She didn't know if she was coming or going. This was obviously the work of someone who really didn't care, and probably used whatever left-over eyes were lying about. Better to offer her for sale without eyes than with mismatched ones.

Eye Repair Problems in Sleep Eyes

The repair of sleep eyes in antique dolls apparently poses more difficulty to the repairer, or at least it would seem that way. It is one area where a great deal of "malpractice" is seen.

Sleep eye repair needn't be so scary, but it does take a little determination and time.

As with anything you would tackle, it helps to know proper methods.

Evaluating Sleep Eye Problems

Eyes That Have Fallen into the Head

This is a common and simple-to-correct problem, if the eyes have broken from the rocker in large clean breaks and the pieces are large enough to re-glue. Clean all of the pieces with isopropyl alcohol and re-glue into place on the rocker with Tacky. Check for proper fit inside the head as soon as glue has set up, in case adjustments have to be made. Allow to dry overnight, soak loose one of the plaster "cups" and re-install the eyes to open and close.

If many pieces are tiny chips or key pieces are missing, the choice is simple: the eyes must be replaced. If some pieces are missing around the back of the eyes, but the front and iris are present, they can be saved by gluing up what is available and building up the interior with plaster, gradually, until you have a round shape. This is a time-consuming and tricky task, but worth it!

Eyes Remain Open when the Doll is Layed Down

In this case, the eyes may be set fixed, or (very commonly found) they may be held in place with a wad of tissue, newspaper, or old rags (I have even found chewing gum!)

Possibly (we can hope!) they are simply stuck in place, and can be gently loosened either manually or with the (lightly applied) heat of a hair dryer.

If the eyes are set fixed and the "rocker" is missing, you might still be able to convert them back to sleep eyes if they can be soaked loose and are intact otherwise.

If the eyes are on a rocker and merely held in place by paper or rags, they can possibly be reinstalled as sleep eyes. Do they fit the eye openings well? If they curve inward too much (look cross-eyed), it may be possible to saw them (gently!) off the rocker, soak loose from the bonding material, rebuild a new rocker, and reinstall them. A lot of work, but worth it.

If they look intact, but are tightly closed, it's perhaps a case of dirt and melted wax (old doll eyelids were painted a flesh color and then dipped in wax, which degrades and melts over the years to make a mess). Clean the inside of the head, using very little water/detergent on a swab, then gently pass a hairdryer. Gently "rock" until you get some movement.

Waxed-over eyelids that are very dirty, but otherwise in good condition can be cleaned with a solvent which will dissolve some of the wax (and the dirt with it). Rubbing alcohol or acetone may be used sparingly (a few drops on a cotton swab) for this task.

If the eyelid paint/wax is peeling away, the eyes need to come out of the head, be thoroughly cleaned, paint/wax removed, then repainted by the dip method, then reinstalled. A bit of work, but really worth it.

It is at this point that it must be acknowledged that this is a lot of material to absorb. If you have never worked on doll eyes and are reading this, you are probably ready to skip this chapter or chuck the whole book.

Take courage. None of the people who repair eyes were born understanding or knowing how to do these many different kinds of steps that are required. The next best thing to do is start looking inside doll heads and observing how the eyes are put in. Expect eyes to take more than one attempt before they look "right". And above all, persevere in this endeavor! Your dolls deserve it.

Materials Needed for Eye Repair

Plaster or other eye-setting compound

Vaseline or Tincture of Green Soap

Isopropyl alcohol

"Sticky" eye wax

Magnifiers

Tweezers

Tacky glue

Replacement eyes

Eyelashes

Hairdryer

Some Eye Repairs Are Simple

Is it a simple case of realigning and regluing the eyes back onto the rocker? If this is the case gently clean the pieces, rinse the glass parts lightly with isopropyl alcohol (rubbing alcohol) and allow to dry for a few minutes. The alcohol not only removes any traces of possible wax from the lids, but is water soluble and evaporates quickly, so it is a good drying agent. It does dissolve some paints, so using it on the body would not be recommended.

Apply a very thin bead of Tacky to both surfaces and gently press together. It will set up rapidly; after about 30 seconds, gently clean away any excess with a wet cotton swab and set the eyes aside to dry for a couple of hours. Once dry, set the eyes to open and close, using the original plaster "pivot cups" if available. If both are still in the head, one of these must be soaked loose with water. Then position the eyes properly and check for fit; then, apply a thin coat of tacky glue to the plaster piece and set in alongside the eye in its original position. Check the operation of the eye now, as this glue sets up quickly. If the fit is good and the eyes operate well, apply a tiny amount of glue around the plaster edge to assure that it will be glued firmly in place.

Lay dolly on her face for a nap of several hours for the glue to dry completely.

Evaluating Eye Repair Tasks

The following photos portray a typical "eye patient".

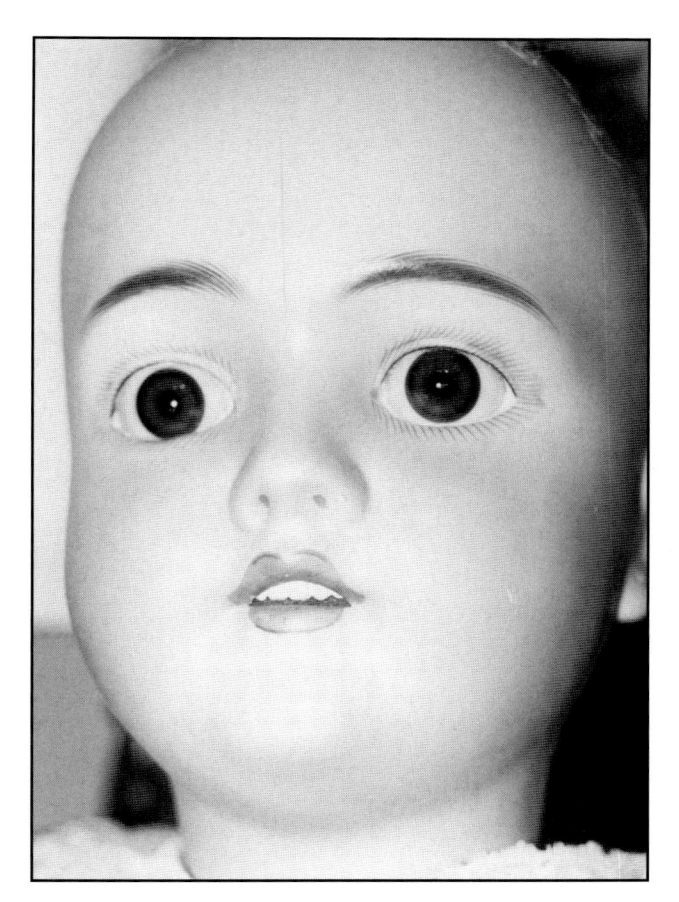

A case of doll malpractice! This beautiful antique "Walkure" head has eyes which are a size too small and are set in crookedly!

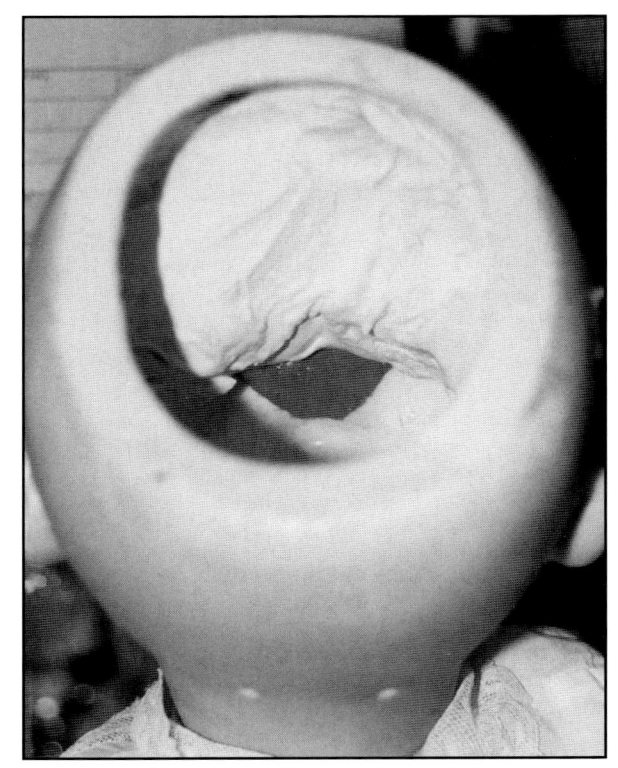

The eyes cannot be removed, even with hours of soaking, because grout was used in the setting. Don't let this happen to your doll!

67

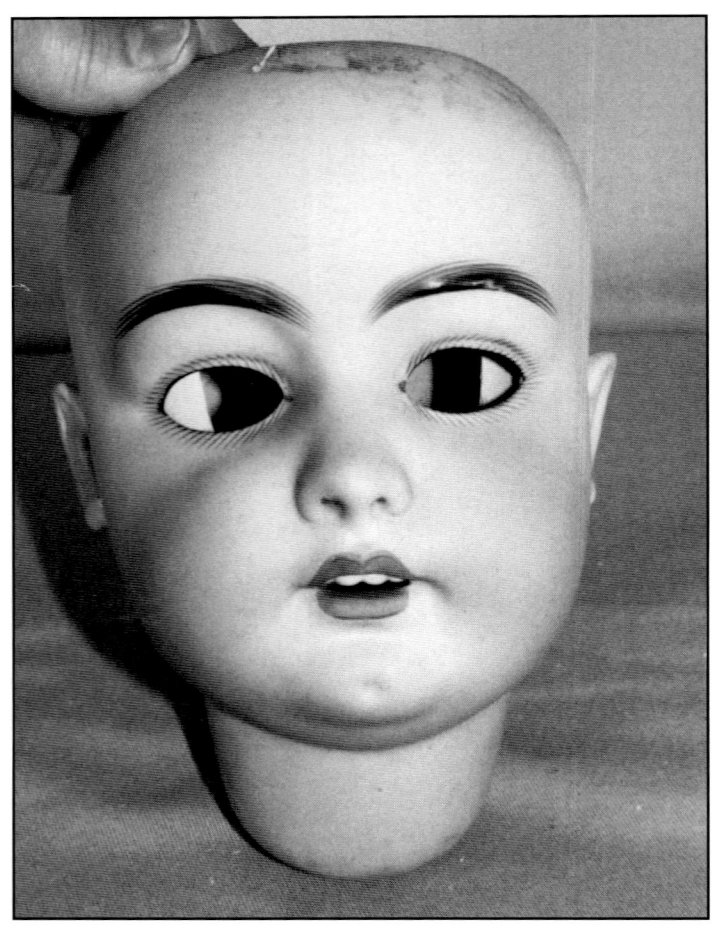

This head, minus eyes, belongs to a fabulous 37-inch Simon and Halbig #1079. Eyes are out and broken, but salvageable.

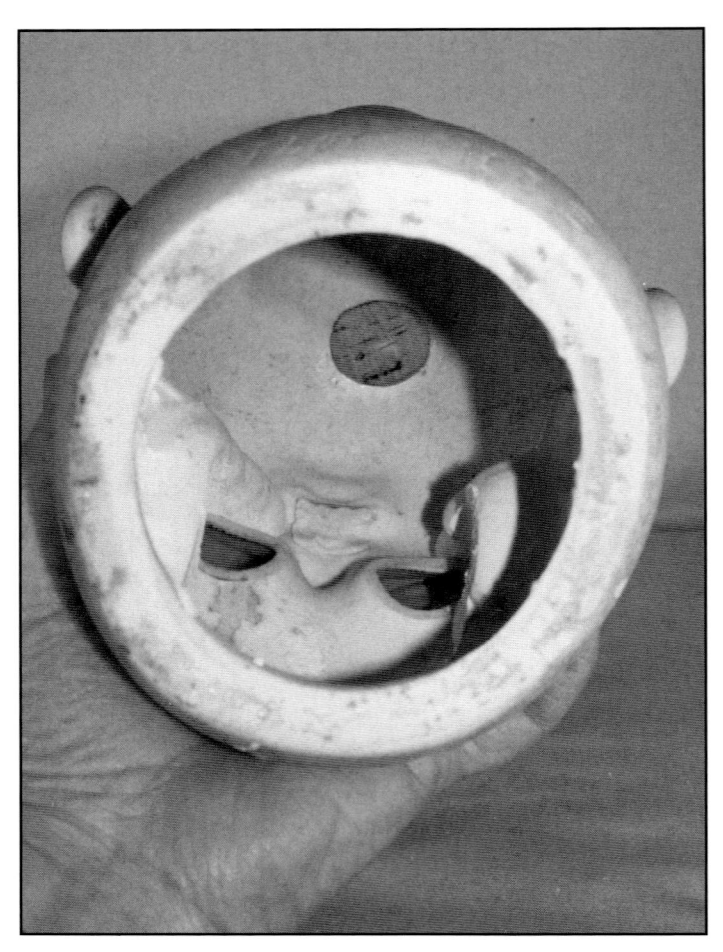

Inside of head, showing plaster eye pivots, in need of a thorough cleaning.

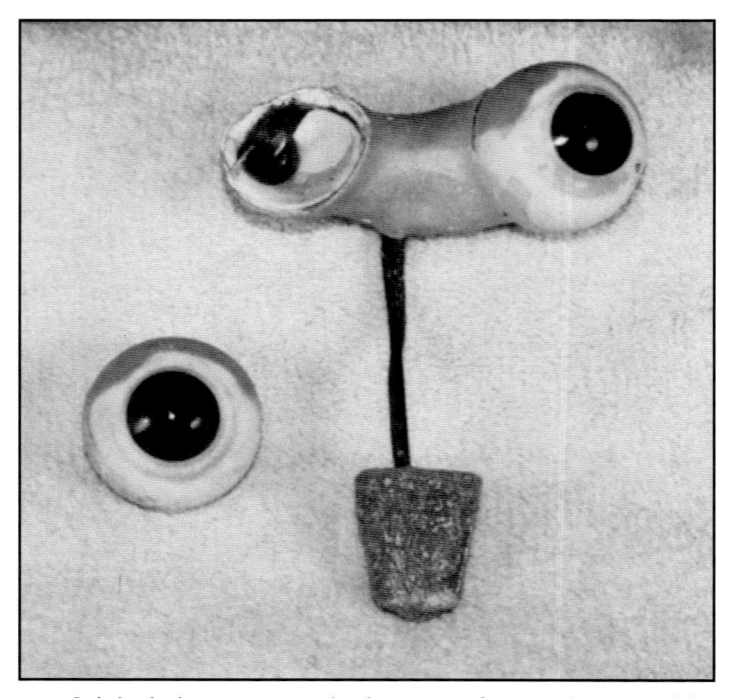

Original glass eyes, one broken away from rocker assembly.

Repainting Eyelids of Glass Sleep Eyes

Eyelids that have wax and/or paint peeling away are less than suitable on lovely antique dolls. The repainting of the eyelids should not be attempted with the eyes still in the head. The old paint and wax needs to be removed and the eyes thoroughly cleaned to get an acceptable result. It is fairly simple to remove the eyes from the head.

First, remove the wig and pate from the head, preferably by soaking, so that the wig will not be torn. In almost all cases, wigs that are on dolls of this age are badly in need of shampooing anyway. The wig can be soaking in several washes while the work on the eyes is being done.

To remove the eyes from the head, lay the head to one side, cushioned and supported by terry towels. Spray a small amount of water around the plaster "cup" that you are going to remove and allow it to soak for a few minutes. Then, using a spatula or other similar flexible, thin blade, gently poke and pry about the plaster edge. It will usually pop up easily. If it resists, allow a few more minutes of soaking, adding a little more water if it has been absorbed by the plaster. Lightly scoring the plaster around its edge away from the eye aids the absorption of the water, and the plaster part will release.

Save the plaster "cup" for reinstallation of the eyes. Now, carefully remove the eyes from the head. You may want to clean the head thoroughly as well.

Clean the paint and wax from the glass portions of the eyes, being careful not to jar them loose from the rocker. Should this happen, they can be glued back into position, taking care that they are aligned as originally positioned.

It may be necessary to use an X-acto or similar sharp tool to scrape away the eyelid paint. Do the final cleanup with isopropyl alcohol and allow to dry.

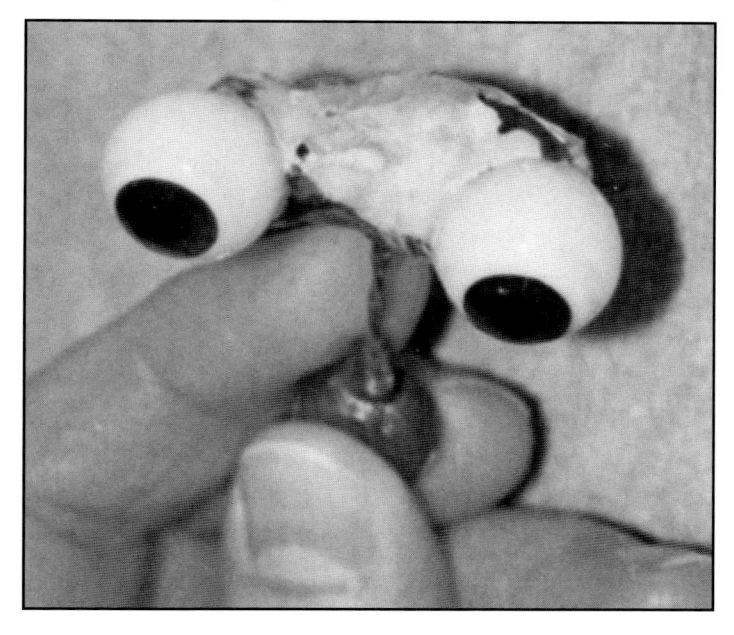

Prepare a small shallow dish with fleshtone acrylic paint and acrylic waterbase varnish, mixed 1:1. Stir this paint thoroughly, but slowly, to avoid any air bubbles. Air bubbles which do arise should be "combed" to one side with a spoon.

With a sheet of wax paper (and of course a dampened paper towel) at the ready, invert the eye assembly, weight up, and dip the eyes into the paint until the paint almost reaches the irises.

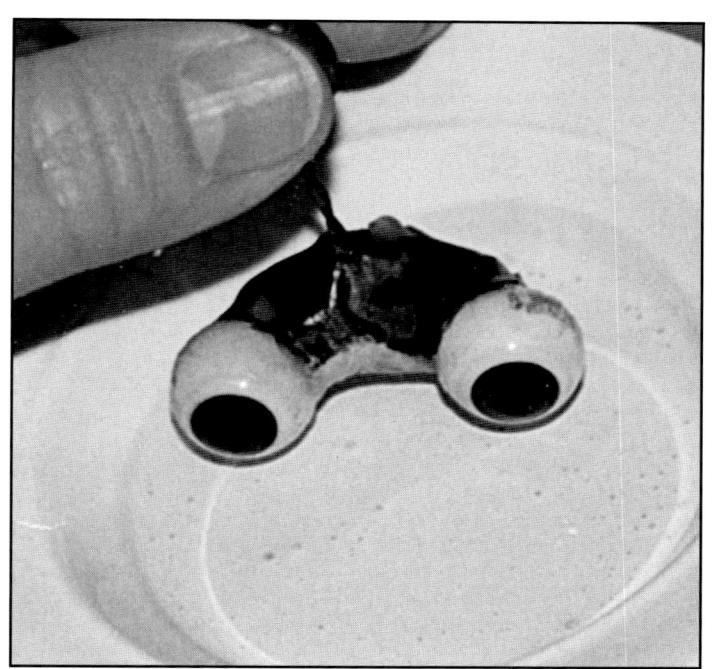

If you dip the eyes too far into the paint, you can always rinse off the paint (detergent, then alcohol) and try again when the eyes are dry. Once this paint dries, it will be much more difficult to remove, so be sure that you are satisfied with your results so far. The aim is to coat the eyes so that the "lids" just touch the top of the irises.

Gently shake the eyes over the paint dish to aid in the flow of the excess paint from the eyes. When the dripping has slowed, place the wet-paint eye assembly on the wax paper, irises up, as in the photo on the following page. As the paint continues to drain, move the eyes a couple of times on the wax paper. Allow overnight drying time.

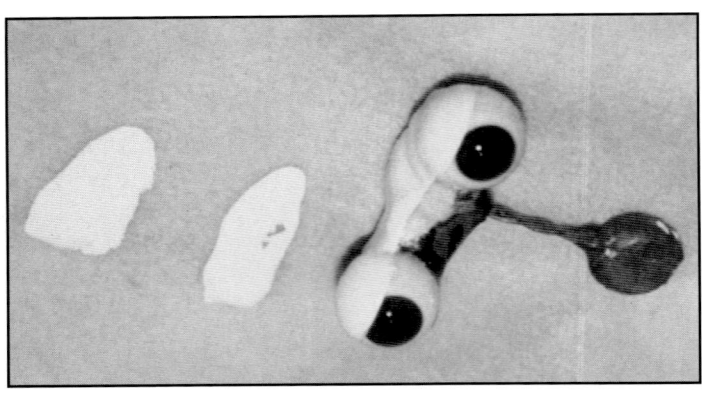

Congratulations! What a professional-looking eyelid paint job you now have! The eyes are now ready to install in the lucky doll's head. See prior instructions for setting sleep eyes.

Setting Eyes Fixed
Case 1: Flat (or Round) Glass Eyes

If the breaks are too numerous or there are obviously chips of eye missing, the eyes probably cannot be saved in their entirety. Is there enough of the iris and front to use them as set-fixed eyes? To most collectors, this would be preferable to replacing them with new eyes

Our subject for this repair session is an antique shoulderplate head, Armand Marseille's *Floradora*. When this head was acquired, not only was it missing the eyes, it also had been broken, and repaired.

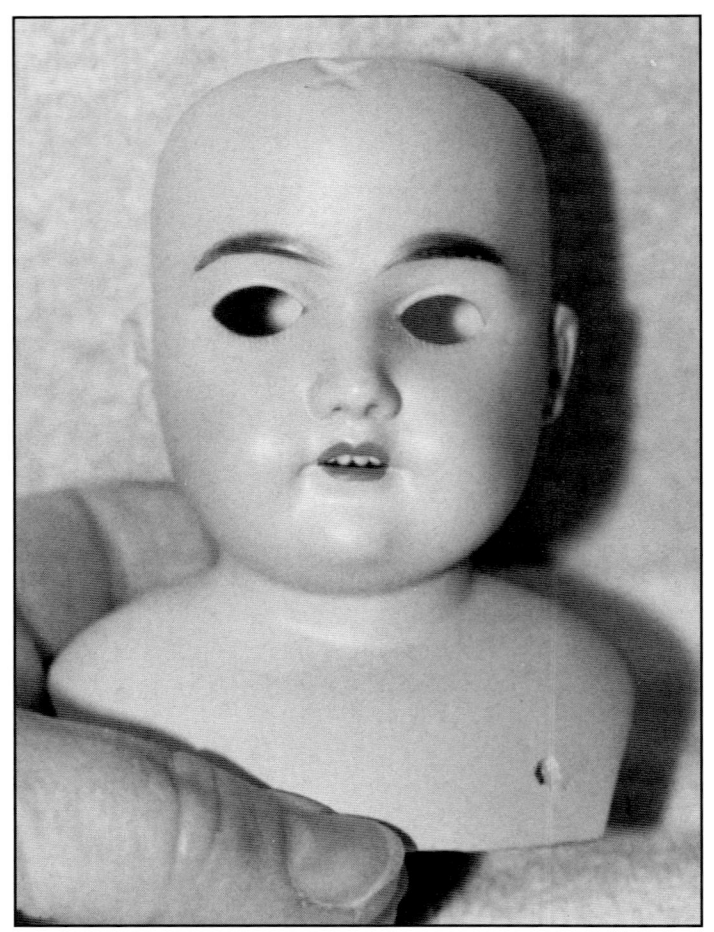

Repair putty is seen in the shoulderplate area and mending grout can be seen inside the head, where the various cracks are located.

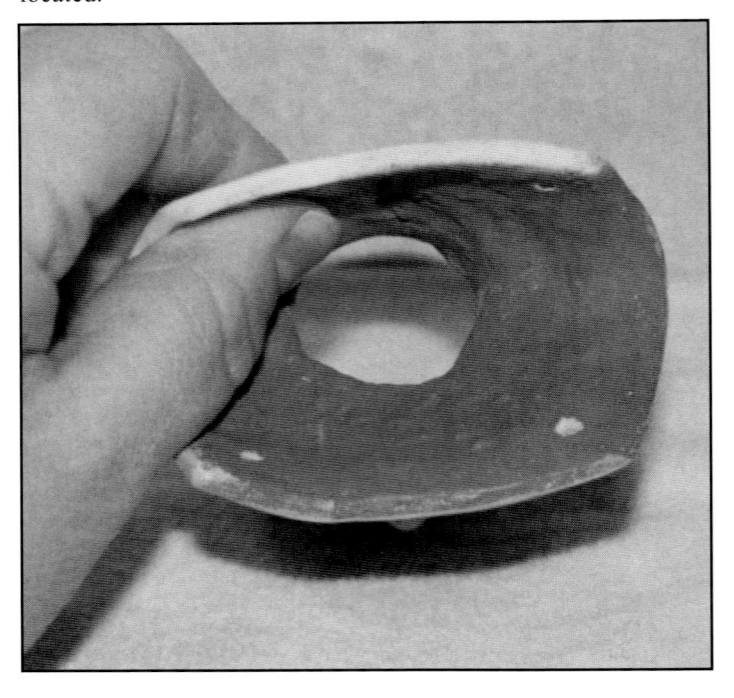

This is an opportune time to mention that if this head were wigged, the head repair would not be noticeable to the average observer. It would be a shame to purchase a doll of this common mark unaware of the repairs, since more perfect examples are

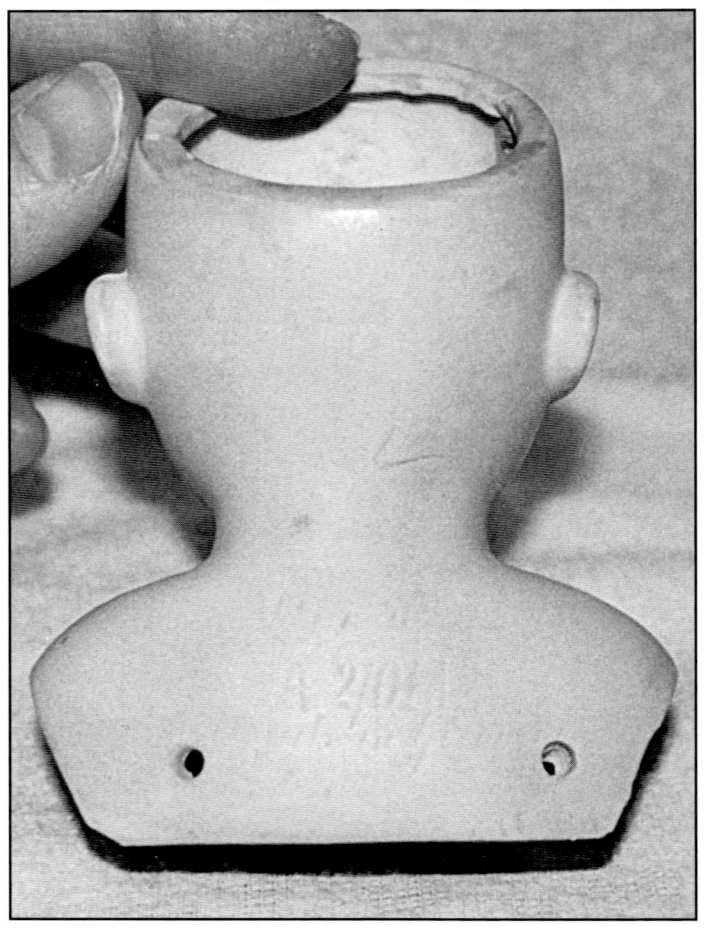

70

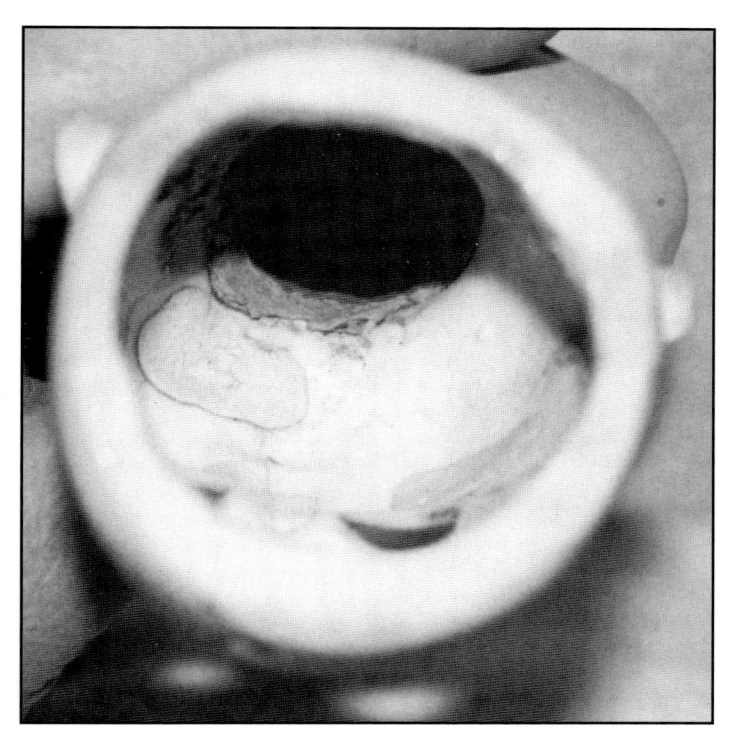

easy to find. This underscores the rule of "If you don't know dolls well, know your dealer".

Since the mends might interfere with the movement of a sleep eye rocker and because the repaired head is of minimal value, the decision is made to install flat glass eyes, fixed.

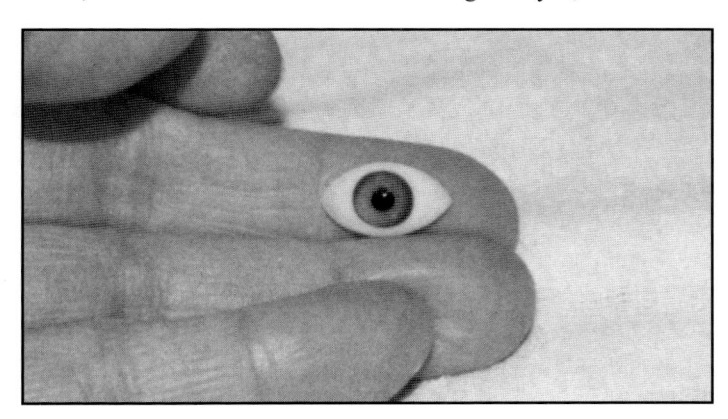

This type of setting can be accomplished by using a thin roll (a "worm") of eye wax, rolling it gently around the front of the eye approximately where it will contact the inside of the eye socket.

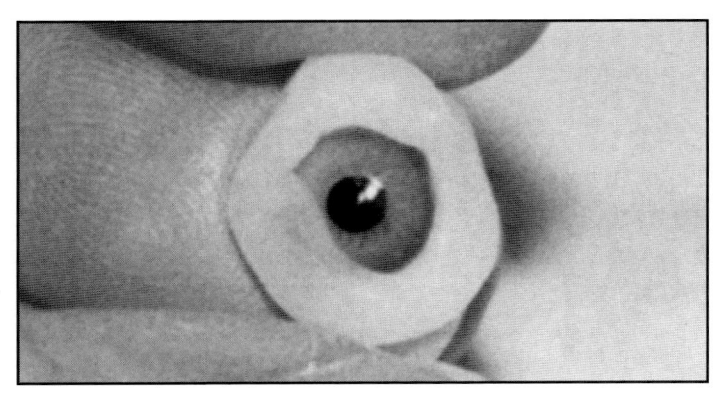

Eye wax (sometimes called sticky wax) is a translucent wax which is pliable to the touch. It is sold in "slabs" or by the individual piece (see below). It is not the old brownish resin which was heated on a burner and used to hold eyes in place temporarily. It helps to warm the sticky wax slightly by rolling it between your fingers or holding it under a work lamp for a couple of minutes. Do not heat this new type of eye wax as, years ago, you would have done with the brown resin, or it will liquify.

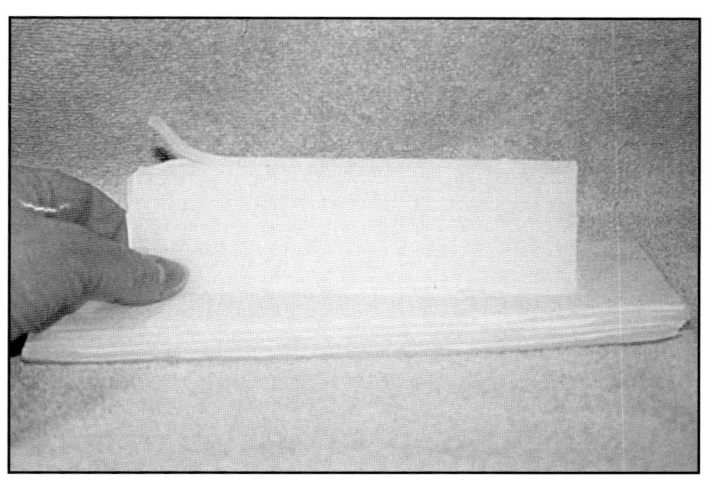

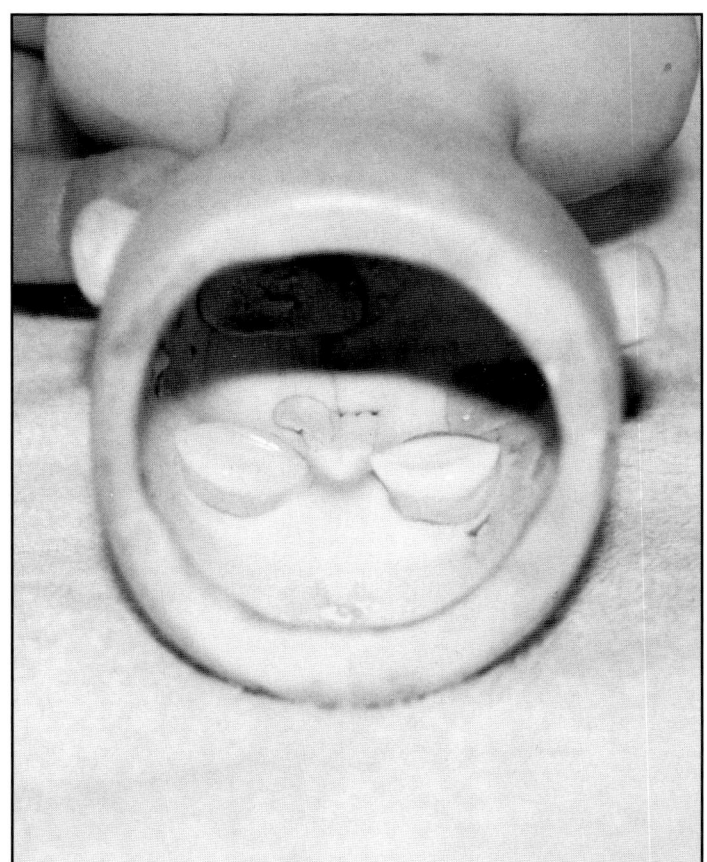

Press the eyes, surrounded with eye wax, gently into place. Using a round wood toothpick, clear away enough wax from

71

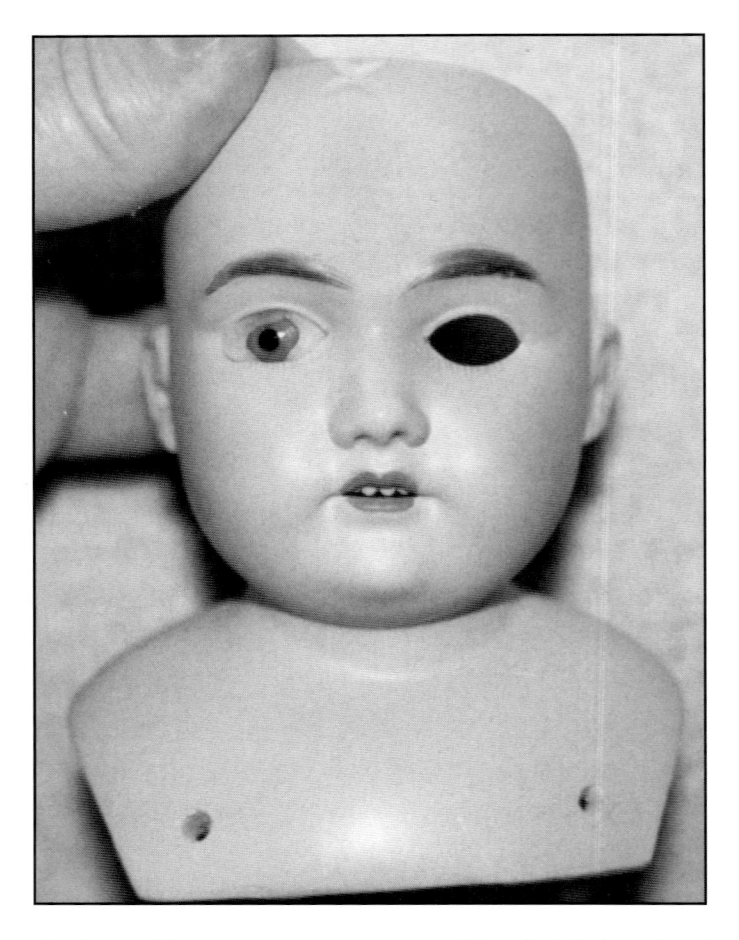

the front of the eye so you can see to position the iris (see above).

Hold the head away and critique the positions of the eyes. The wax will hold them in place but allow them to be adjusted about until they have a pleasant gaze and are approximately focused. "Approximately" may be the best result to be had, as very often antique glass eyes are not nearly identical, nor are the socket openings usually the same (see below).

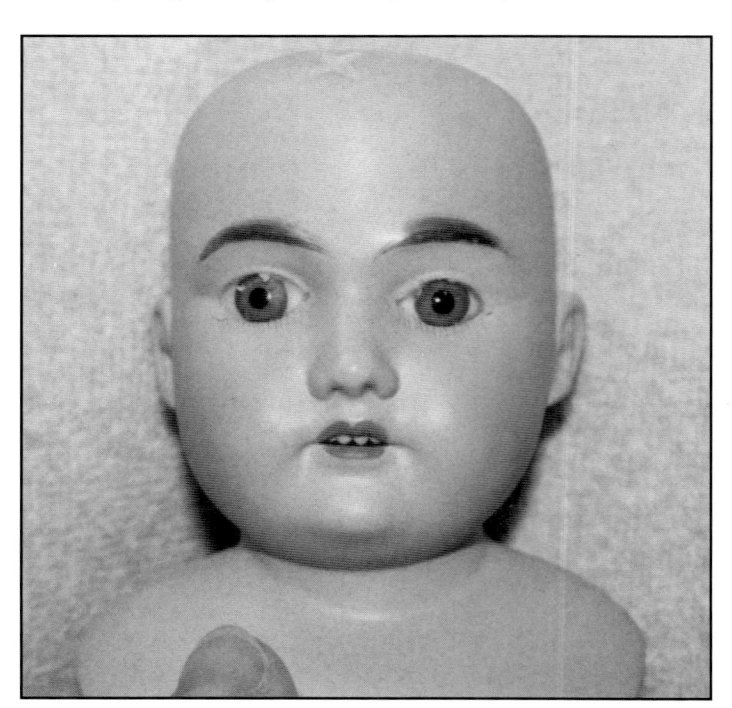

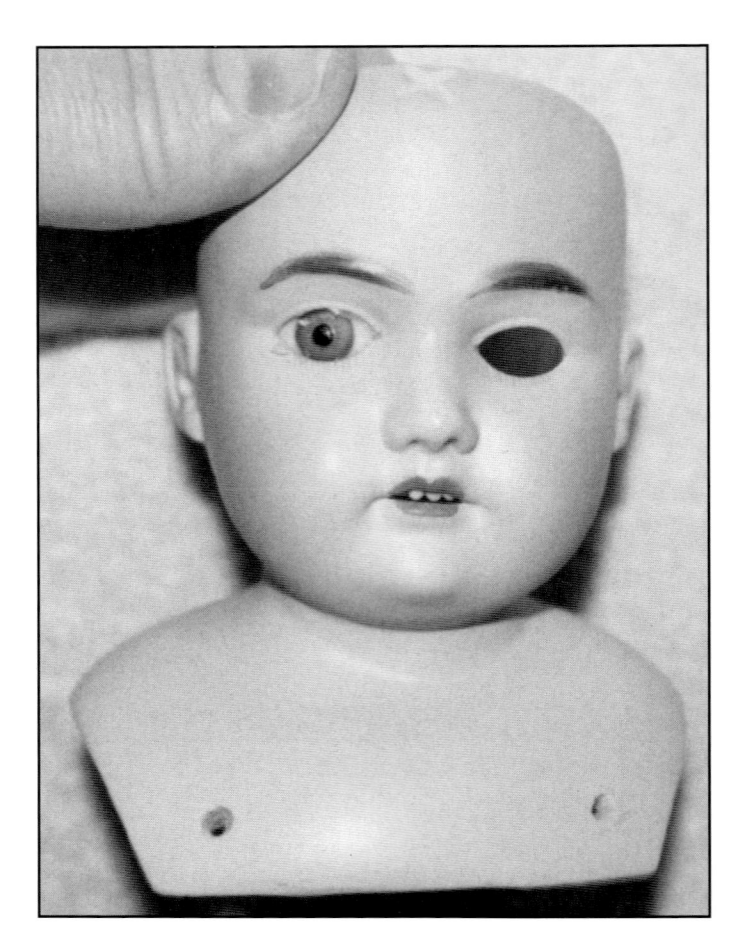

Sometimes switching the eyes gives a better balance, and is worth a try if the eyes seem unusually "out of match". The thing to remember here is that the purpose is to save the doll's original eyes and use a method that is reversible, should some later custodian of the doll have higher skill and more advanced methods become available.

When the restorer is satisfied with the doll's gaze, position the head so it is level and tip it slightly upward at the neck end so that there is no chance of plaster running down to and out of the mouth. Using a small, rolled-up towel is usually all that is needed to hold the head while the eye compound dries. Then apply your choice of eye-setting compound and allow overnight drying time (see top photos, next page).

When the eye compound is dry, clean away any traces of eye wax from the front of the eyes using a wood toothpick, then isopropyl alcohol on a cotton swab.

Now *Floradora* can see, and looks lovely with her new eyes (see bottom left photos, next page).

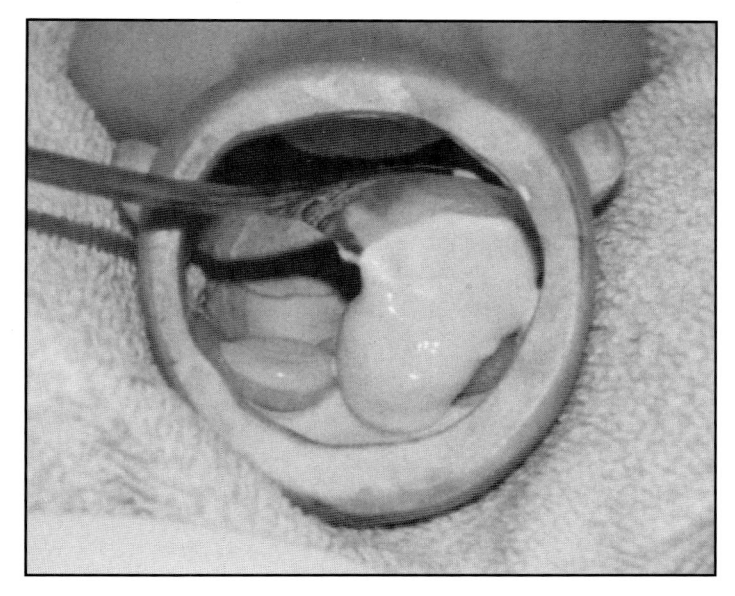

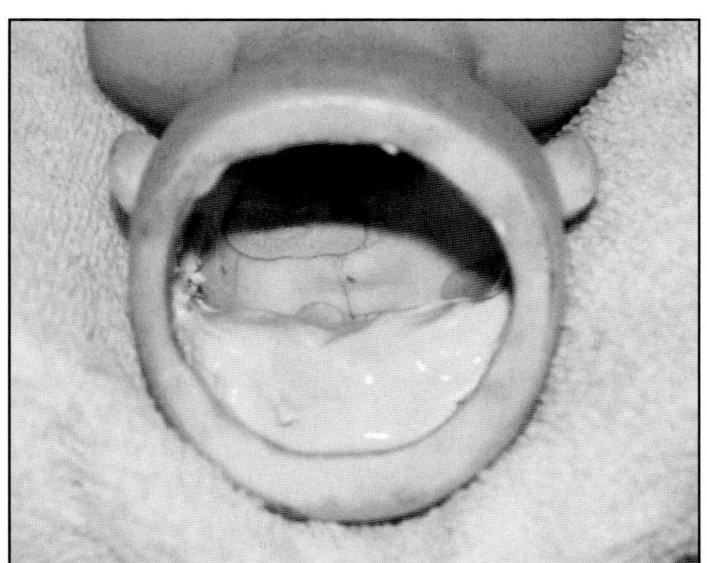

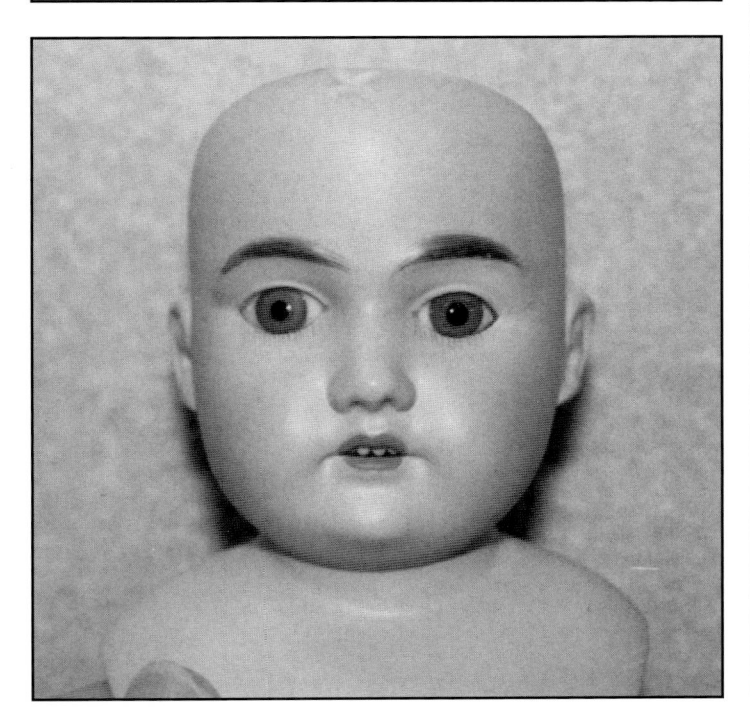

Another gentle warning, relative to setting eyes: many restorers use plaster for this step, and it is not altogether unsuitable; others prefer to use the newer compounds which do not heat up with the addition of water and do not expand. Plaster has been widely used for scores of years and is probably acceptable in most uses of this type. The critical thing to remember here is to use a compound which can be soaked loose in warm water, if necessary. DO NOT USE GROUT for setting doll eyes, as it is not removable. Sometimes the home handyman or woman must be kept at arm's length! One of the rules of doll repair is that not everything under the kitchen sink or in the basement workroom is suitable, though that is where many useful products are "discovered"!

Setting Eyes Fixed
Case 2: Paperweight Eyes

Because the techniques will be virtually the same as for the flat glass eyes above, we will not elaborate on each step as thoroughly. Do read through the above section if you will be doing this activity.

Our candidate for this repair session is a new (reproduction) but beautiful French A.T (see below). We just don't have an antique French head rolling about, at the moment, neeeding replacement eyes!

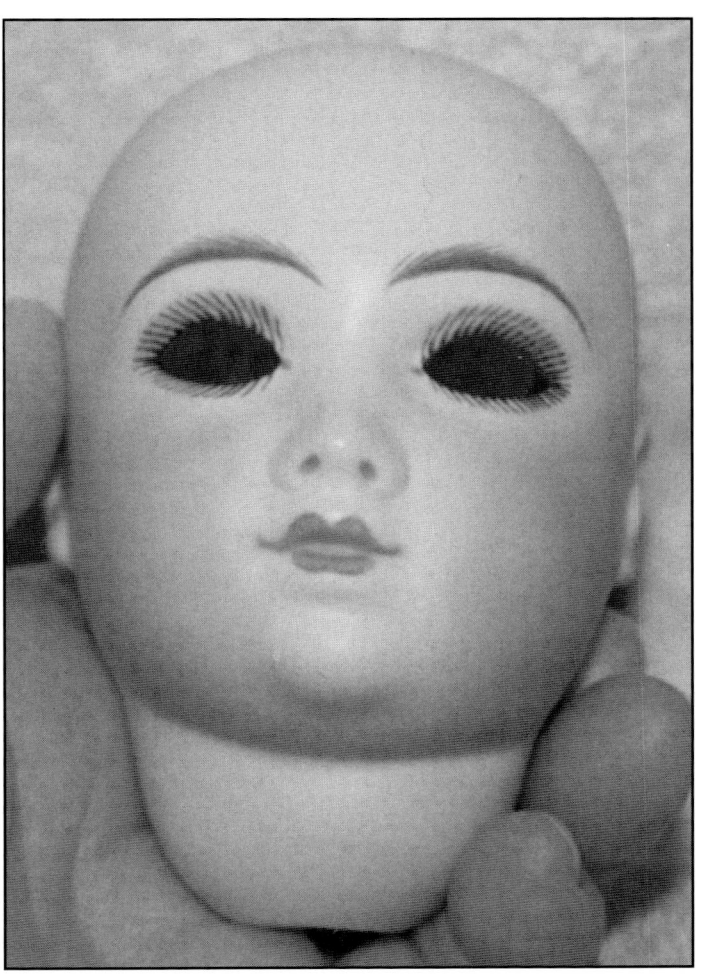

This type of head requires paperweight eyes. When the eye openings were cut and sized, they were slanted to accept the paperweight shape. We opt to use a beautiful glass pair rather than the inexpensive acrylic "paperweight" eyes that are on the market. They don't compare to the quality glass eyes that are available.

Viewing the paperweight eye in profile shows clearly that this type of eye has a prominent bulge and could never be set "to sleep".

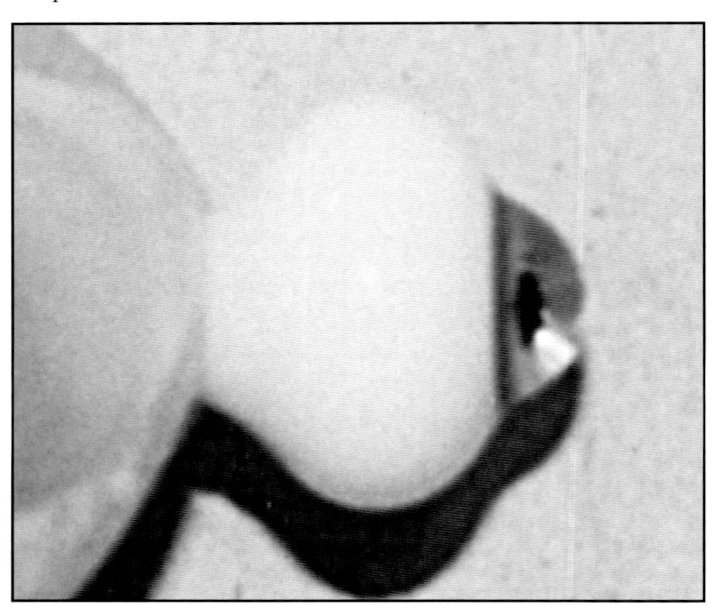

Both eyes set in head with sticky wax.

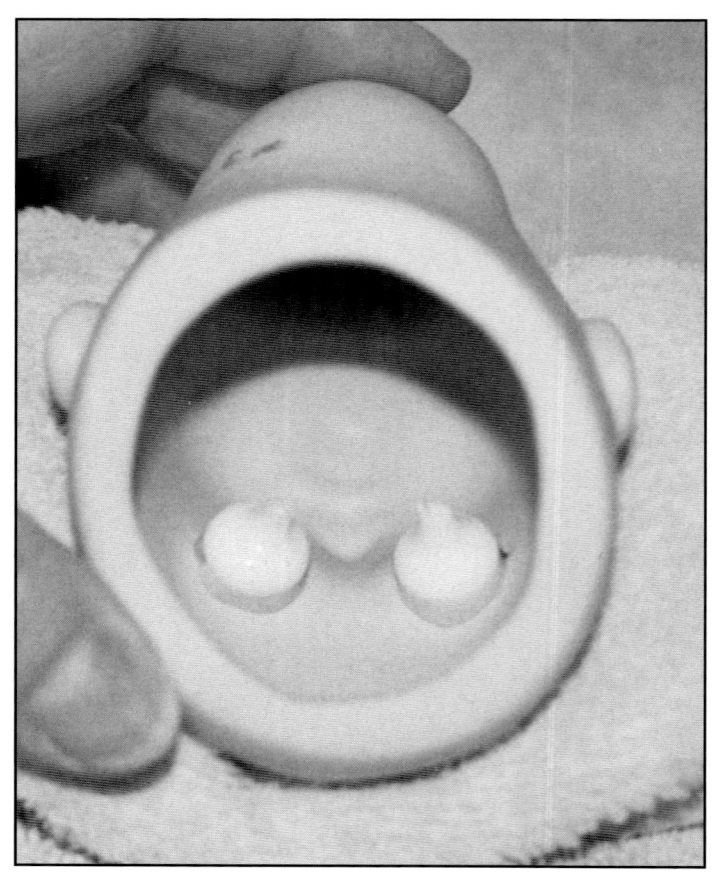

The head propped so that it is level, the plaster applied, and overnight drying time allowed.

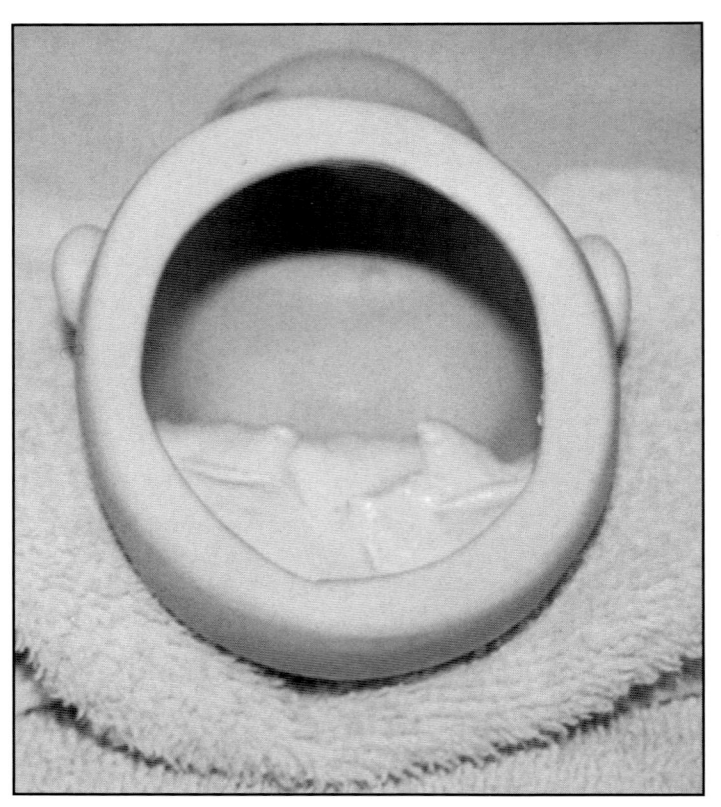

The final result.

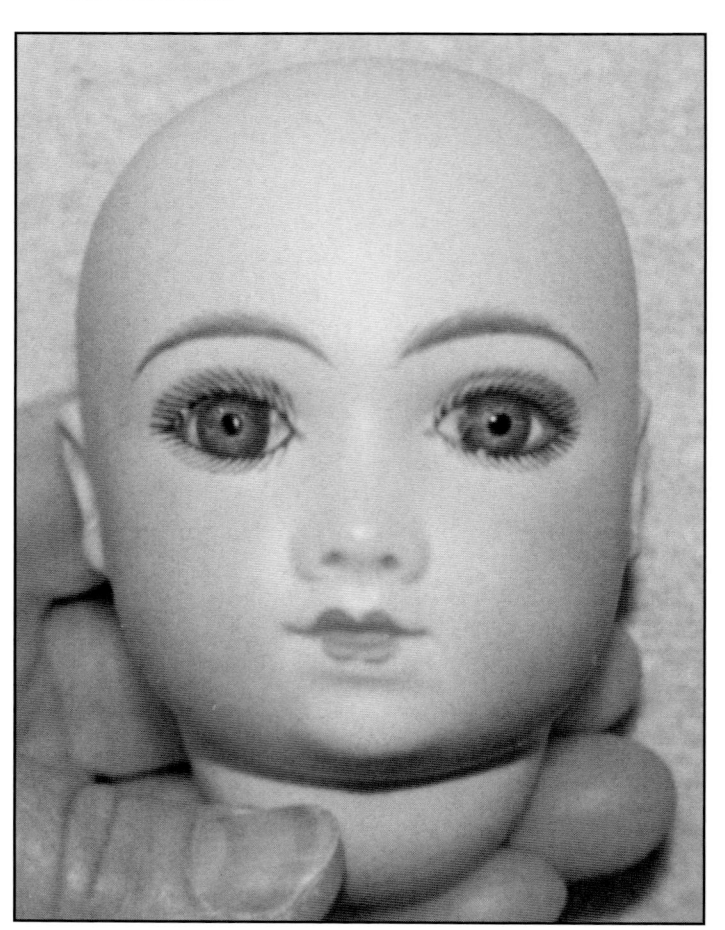

74

Sleep Eyes Out of Head, Not Broken

All repair jobs should look this difficult and be this easy.

This antique Cuno and Otto Dressel head has its original sleep eyes and pivots, present but out of the head.

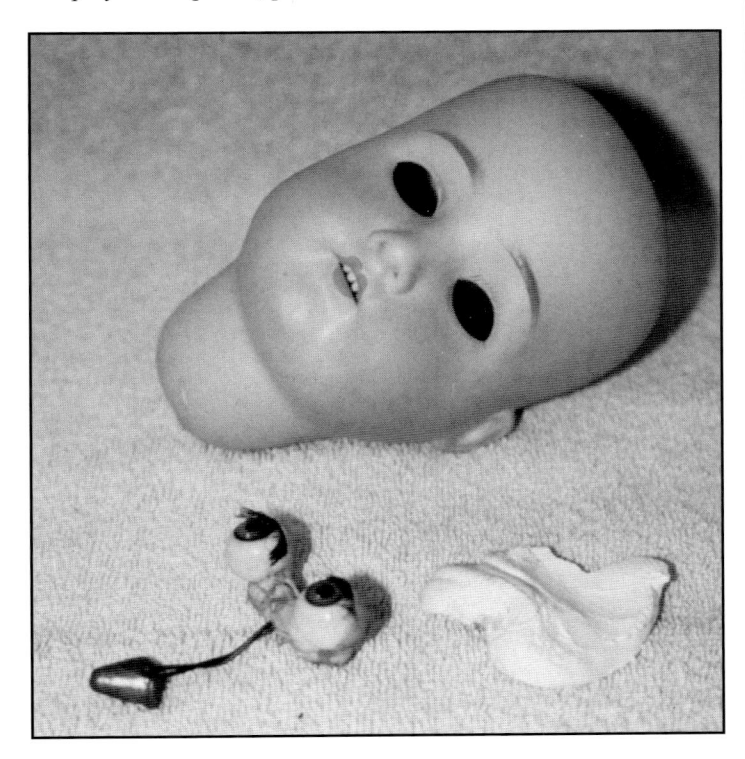

First, the fit of the rocker is checked by simply holding the eyes in position inside the head. The fit and alignment are acceptable.

Glue the original plaster pivots in place, one at a time, with Tacky glue.

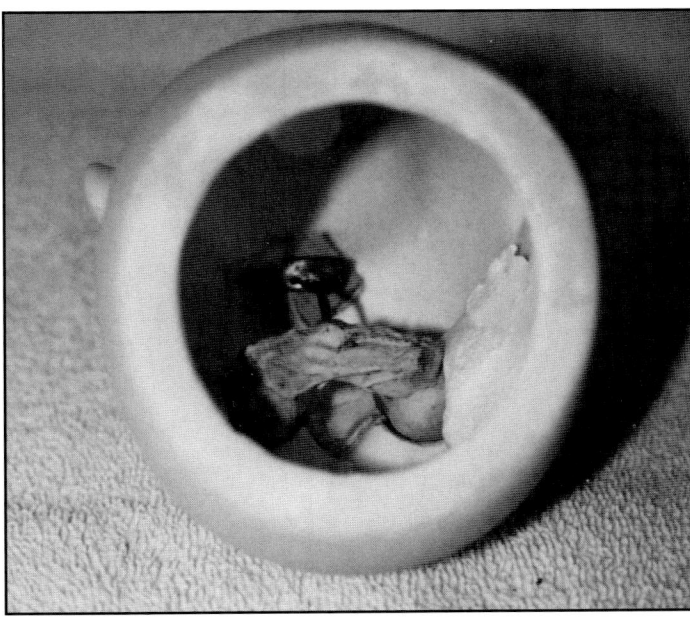

Holding the plaster pivots in place manually, check the rotation of the eyes to assure that they open and close properly. Allow overnight drying time for the glue.

A successful eye repair!

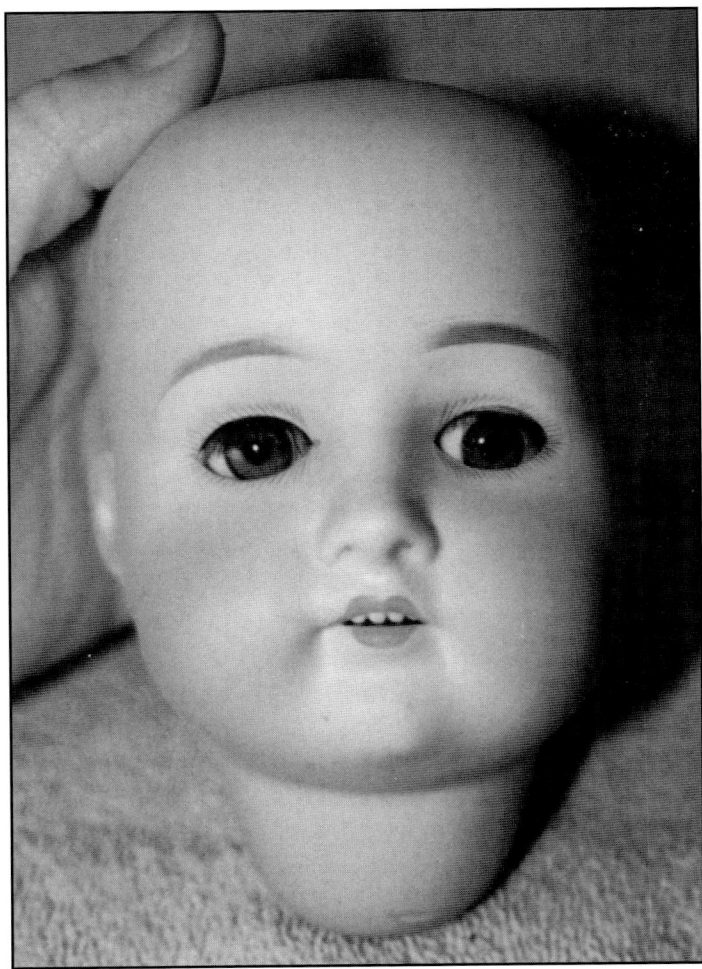

75

Eyes out of the Head, Broken

Since all the pieces of the broken eye are present, this repair task is fairly simple.

The eye to the right in the photograph was off the rocker and broken into several pieces. Easy but time-consuming gluing, using Tacky, gets it all back into place. Drying time must be allowed between the gluing of each piece. The photo shows the next-to-last piece ready to be placed. When dry, the excess glue is cleaned from the surface using isopropyl alcohol. The eyes are then set as sleep eyes, using the procedure given above.

Original Eyes That Are Mis-aligned

This problem occurs for the following reasons:
1. Time and head have caused the rocker to warp, turning the eyes inward.
2. Someone inexperienced worked on the eyes, with inadequate results.
3. Someone put an old set of eyes in the head, and the eyes do not fit (see # 2).

For this section, the example shown is the large Armand Marseille #370 which was shown under the Replacement Body section in Chapter Four. (See top right photo.) This close-up photo shows the serious misalignment of what appear to be original eyes.

It's necessary to remove the wig before the eye problem can be completely and accurately diagnosed. To maintain continuity in this lesson, the head soaking/cleaning is also shown here.

Because the wig is tightly glued on, the head is soaked in a shallow depth of water, enough to loosen the wig but not totally dislodge the eyes, as their condition is unknown (see center photo).

The wig releases easily from the head and the pate can be peeled away from the wig without any damage to the wig cap. Notice the amount of soil in the water and the excessive amount of glue on the pate! (See bottom photo).

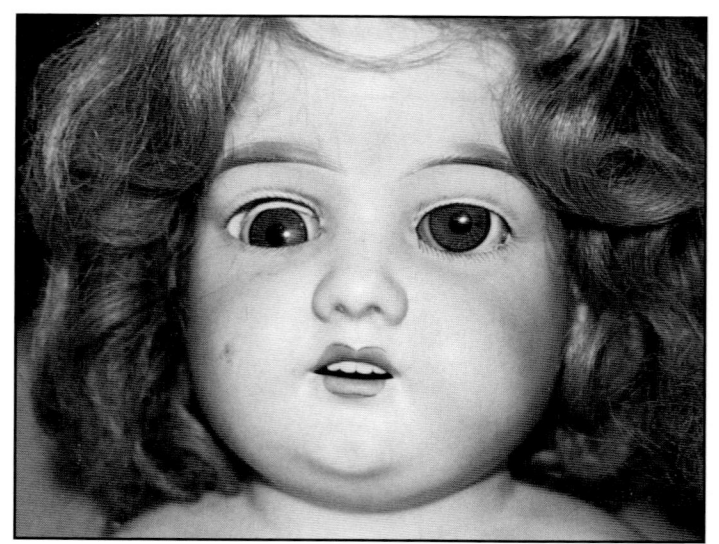

The eye condition can be assessed.

It would appear that melted wax from the eyelids and dirt are the main causes of the problem.

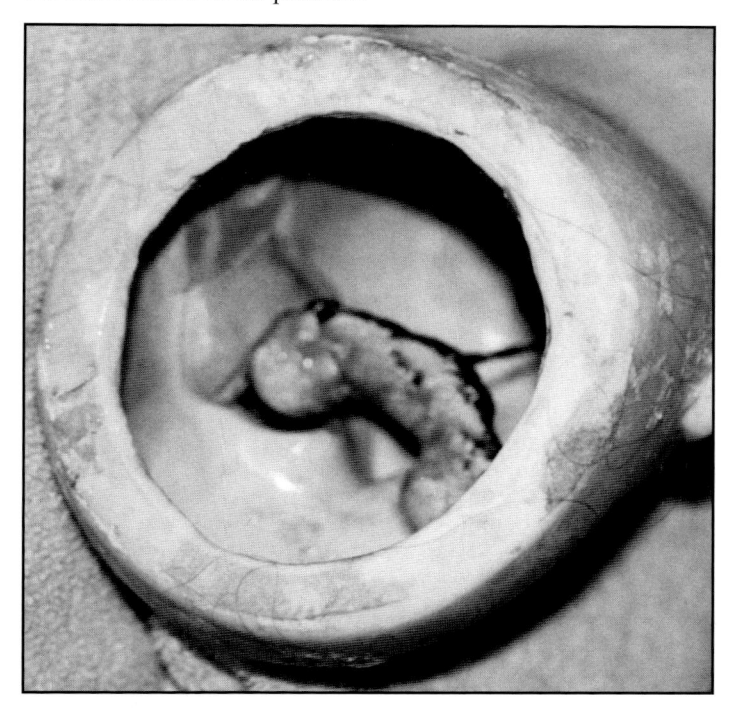

A small amount of water is sprayed onto one of the eye pivots and allowed to soak into the plaster of the pivot for a few minutes.

Using a thin scalpel blade, the pivot is gently prodded until it loosens and comes out, allowing the eye assembly to be removed.

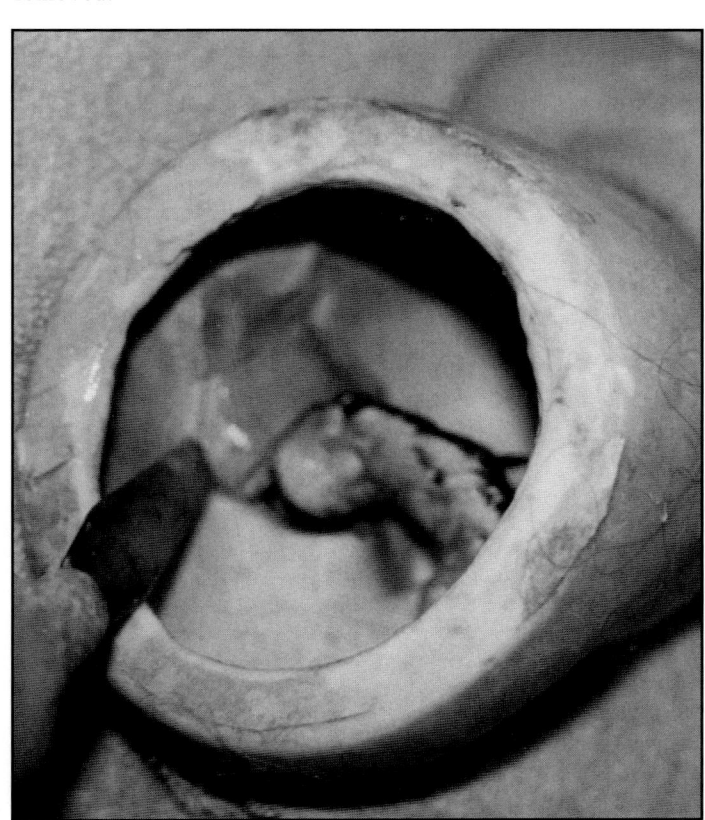

Using an X-acto knife, the excess wax is removed by gently scraping it away. A thorough cleaning with isopropyl alcohol of both the glass eyes and the painted rocker assembly follows.

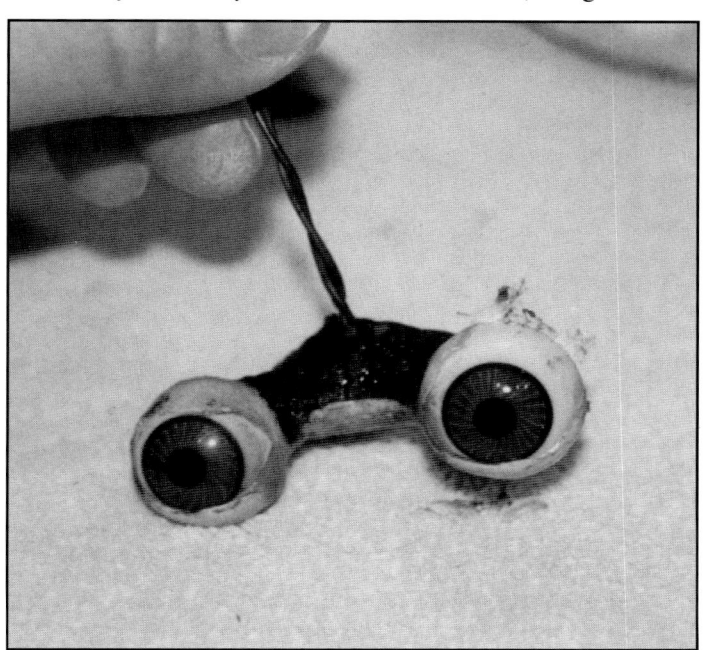

The eye assembly is re-installed in the head, using the same procedure in the following section: "Sleep Eyes Out of Head But Present". How much better the eye assembly fits now. (This proves our rule "Dirt is not a virtue.")

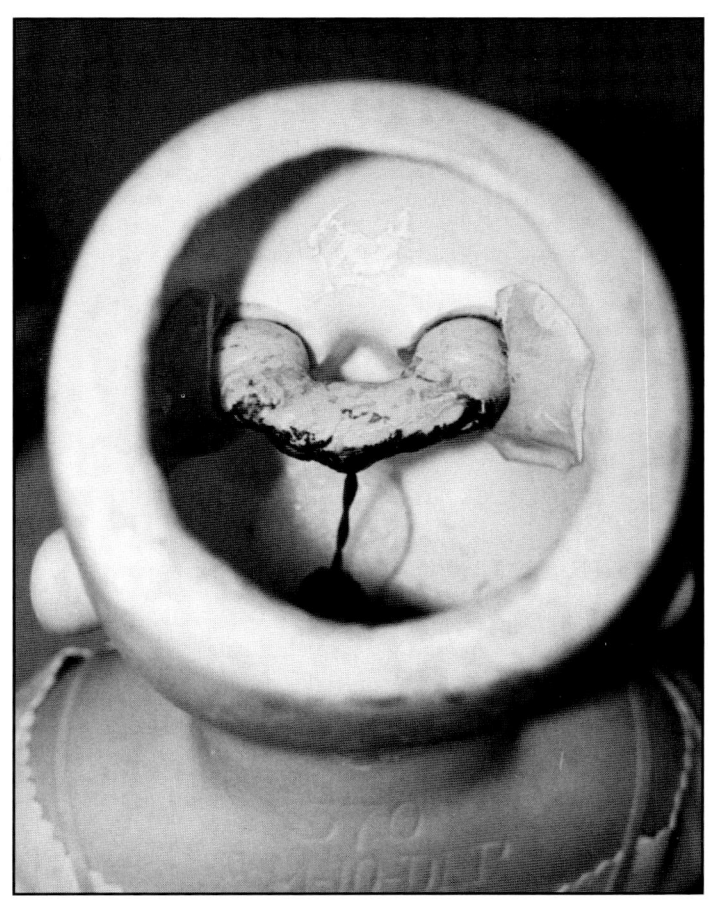

The finished eye work. What a difference from the original photo!

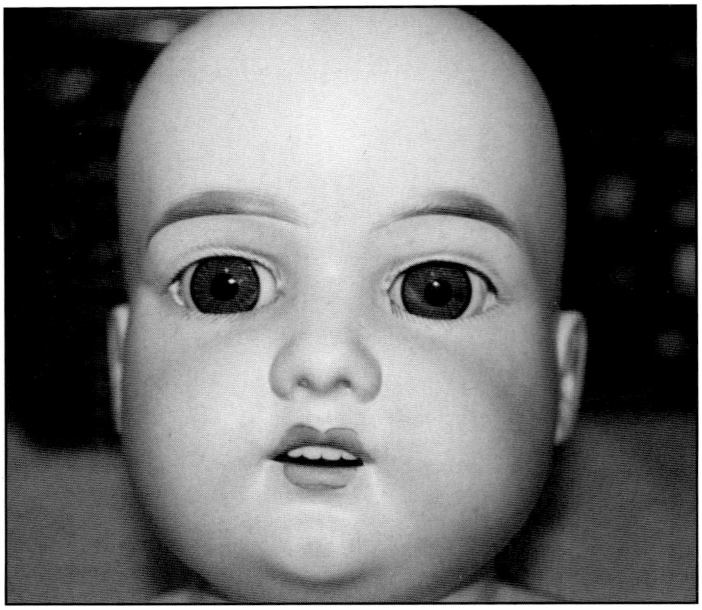

A new pate is glued to the open crown, in preparation for a wig.

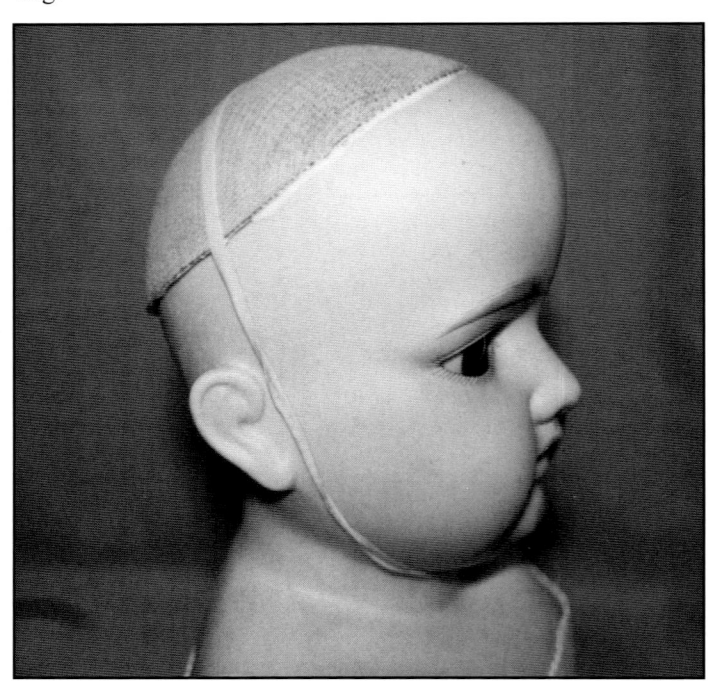

invert the entire head, wig and all, into a shallow pan of warm soapy water (see top left photo, next page).

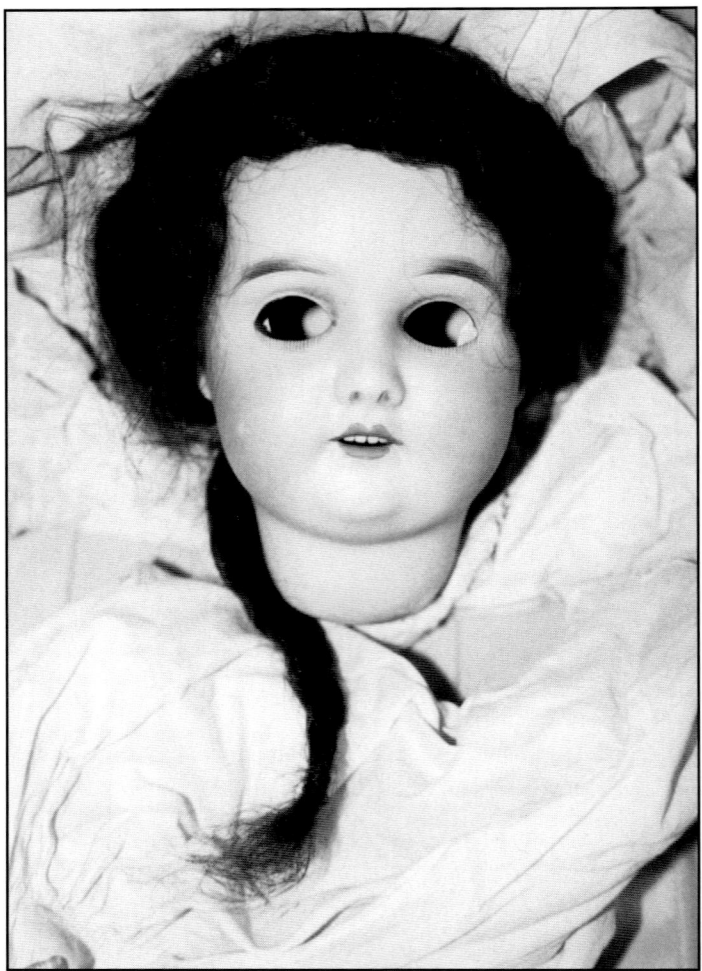

After a few minutes the wig easily peels away from the bisque without tears or "pulls" to the bisque finish. Notice how dirty the water is. The original neck button will be set aside to dry (see center photo, following page).

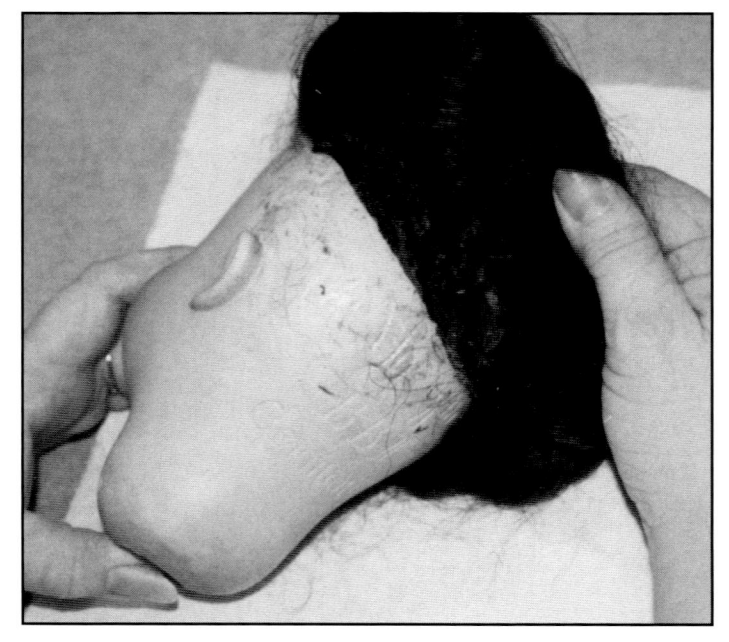

Sleep Eyes Out of Head But Present

Pictured at top right is a George Borgfeldt doll, marked BBII. In this situation, the original eyes have fallen out of the pivot cups, but are still in the doll's head.

The wig is not loose enough to remove without risk of tearing it. We notice that this cute dolly face is one we always love, a CM Bergmann with the additional mark of BBII (see bottom right photo).

Turning gently to avoid further damage to the eyes, we

78

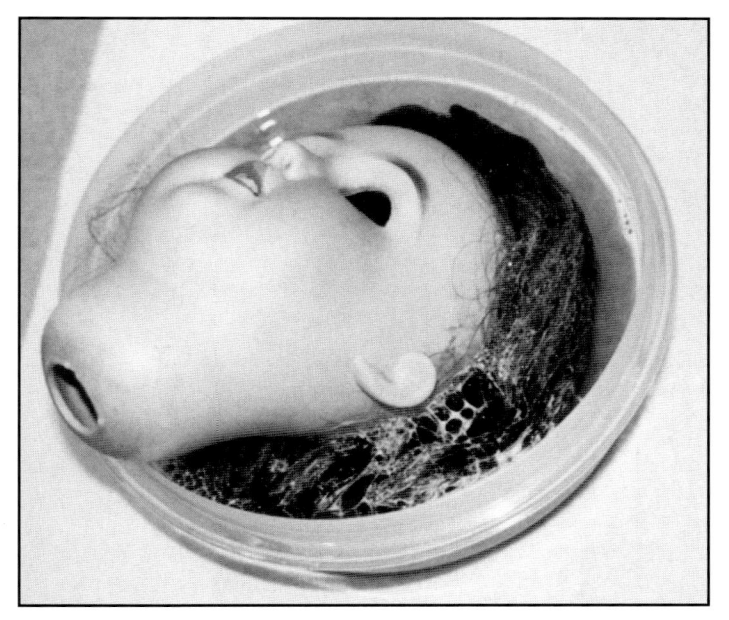

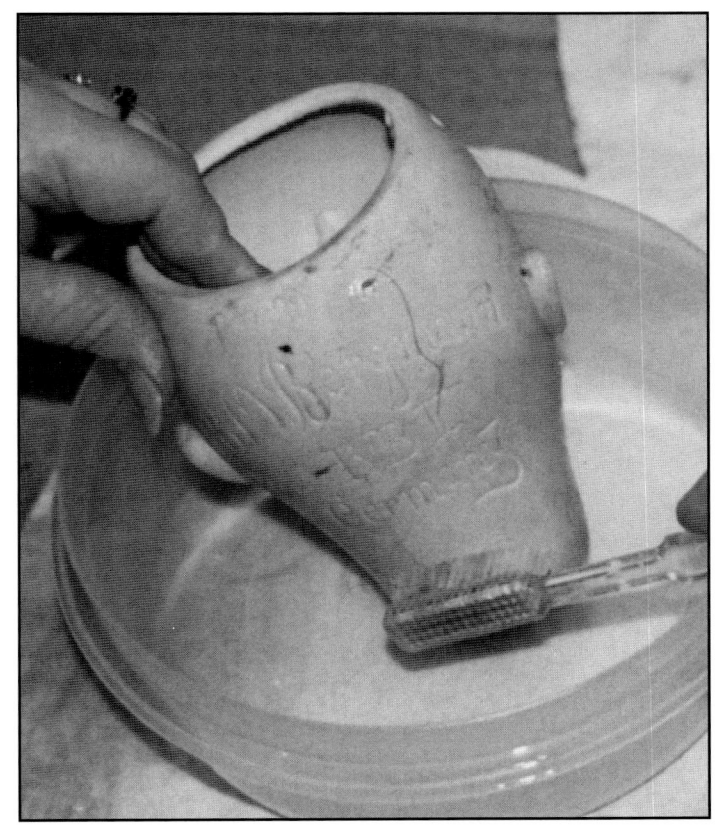

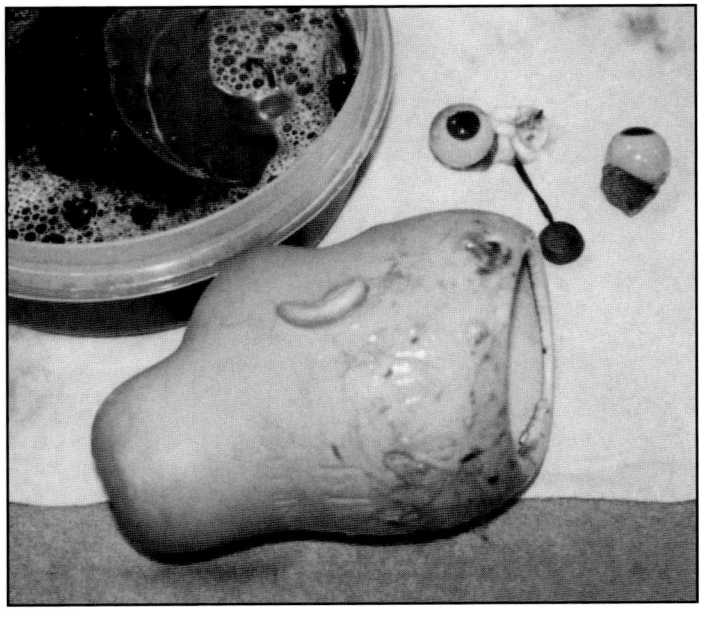

Continue soaking the wig and pate so that it can be removed without tearing the delicate old net backing of the wig. We want to keep this backing as intact as possible. The wig repairs are discussed in Chapter Six, Wig Care. We have carefully retrieved the eyes from the doll's head (see left).

A few water changes later, we are starting to really see some results with the bisque head, as well as the wig. This process so far has taken around forty minutes (see top right).

The head is clean enough now to begin a gentle brushing, inside and out (see above).

We have cleaned the head and removed one of the eye pivots, since we need one out of the head to install the eyes.

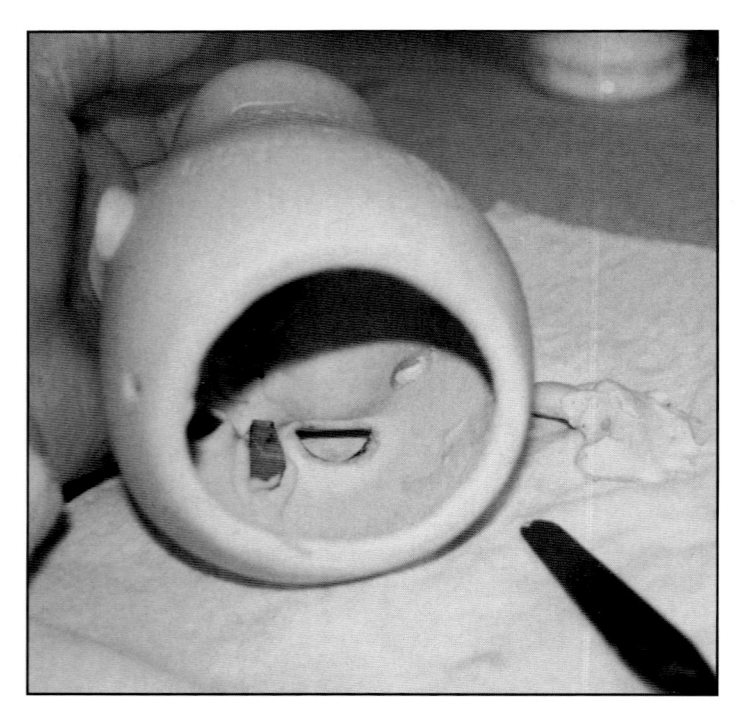

The wig water is finally beginning to be less dirty and the pate has soaked loose.

A trial glue-up of the eye assembly is done and the fit is tried in the head. The eye assembly fits well and the eyes are lined up properly, but the assembly is weak, with old cracks in the rocker material. It's necessary to build up the rocker while in the head, so as not to change the alignment, which would surely happen if it were removed for this work.

The eyes are temporarily "tacked" into the sockets, using a little sticky wax on the front of each eye, to hold the rocker assembly in the desired position while the build-up is done (see top right).

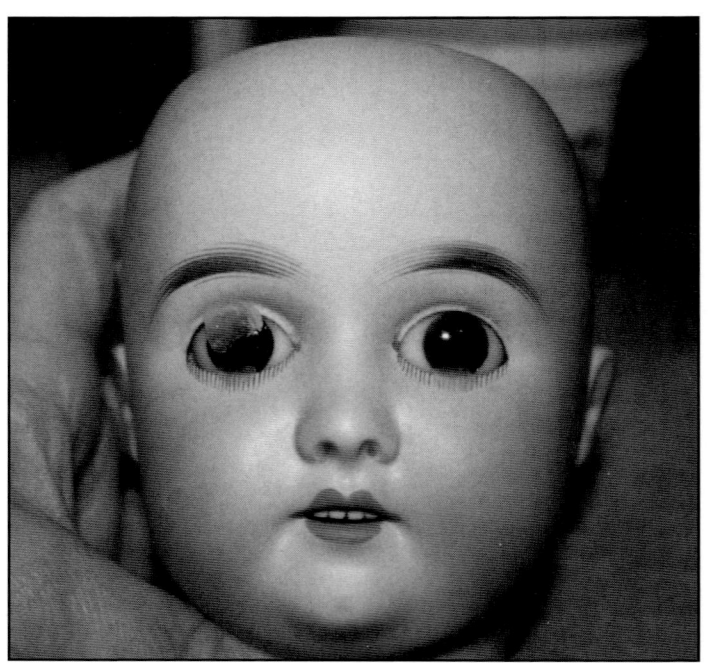

Using a spatula, wood putty is applied, using enough to build up the rocker that the eyes will be stationary when dry. Overnight drying time is allowed (see below).

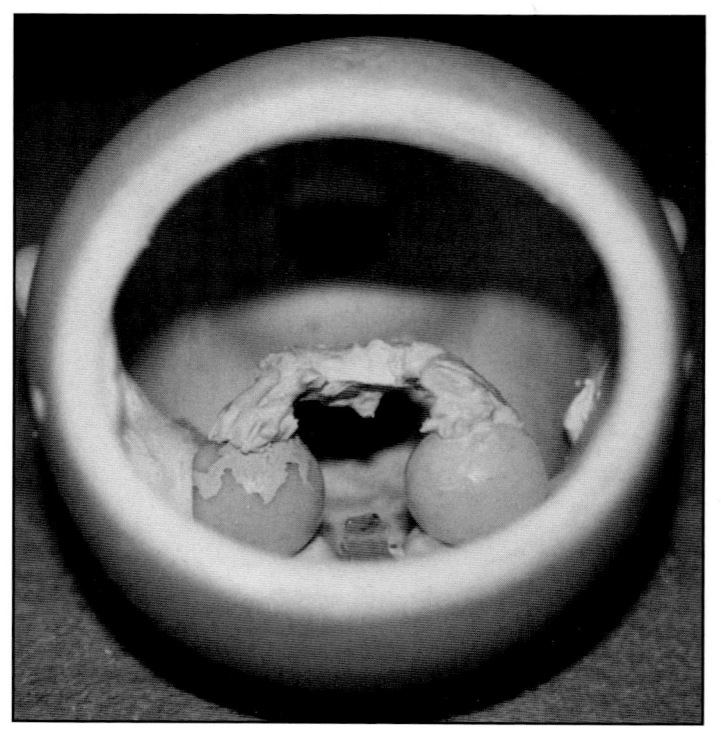

The sticky wax is now removed from the front of the eyes and the rocker assembly is carefully removed from the head.

Now begins the easier task, building up enough putty around the eyes and rocker that it will be a sturdy unit. It's much easier working outside of the head!

The second application of putty is begun, using a wet paint brush to smooth out the putty. Again, overnight drying time is needed.

With the third application of putty, the rocker looks like there is enough putty to securely hold the eyes in place for the doll's future life. A final overnight drying is needed.

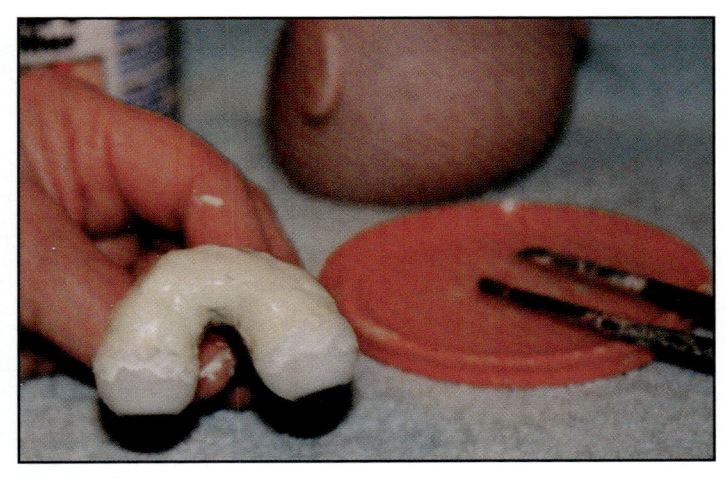

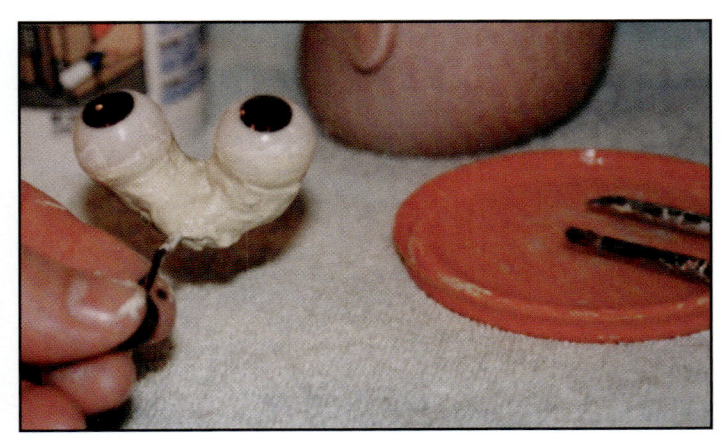

When the putty material is dry, a final cleaning of the glass eyes is done to remove any body oil from handling and to clean away the film of putty at the edges of the work.

The dry putty portion of the rocker is painted by brush, in flesh acrylic. While it is not essential to do this, as it will not show inside of the doll head, it is a matter of completion and pride in the job.

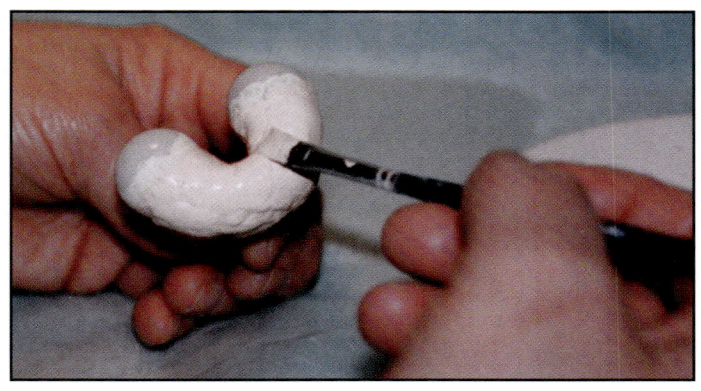

How to Paint Glass Eyelids

Now the eyes are ready to be dipped to produce a professional eyelid painting result.

Use a 1:1 mixture of flesh acrylic paint and waterbase acrylic varnish in a shallow dish. The eye mechanism is dipped just enough that the painted lids will touch upon the irises (see below). Lay the mechanism on wax paper to allow the excess paint to drip away. It can be moved a couple of times in the first few minutes so that it is not sitting in a puddle of paint. Allow overnight drying time.

81

Eye rocker is dry and ready to install.

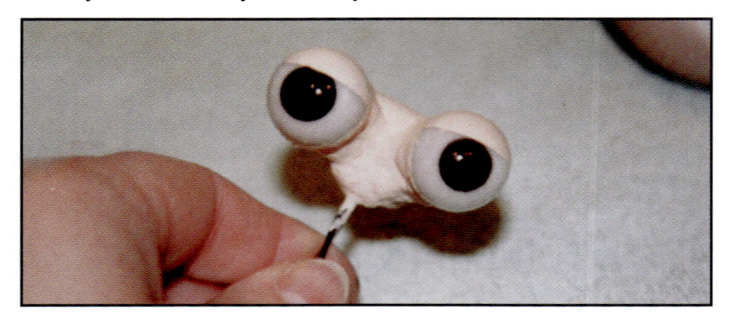

Now the eyes will be re-mounted in the head. With one plaster pivot holding the rocker in position, check the appearance of the eyes from the front. Tack each eye in place with a small amount of sticky wax on the front of the doll's face.

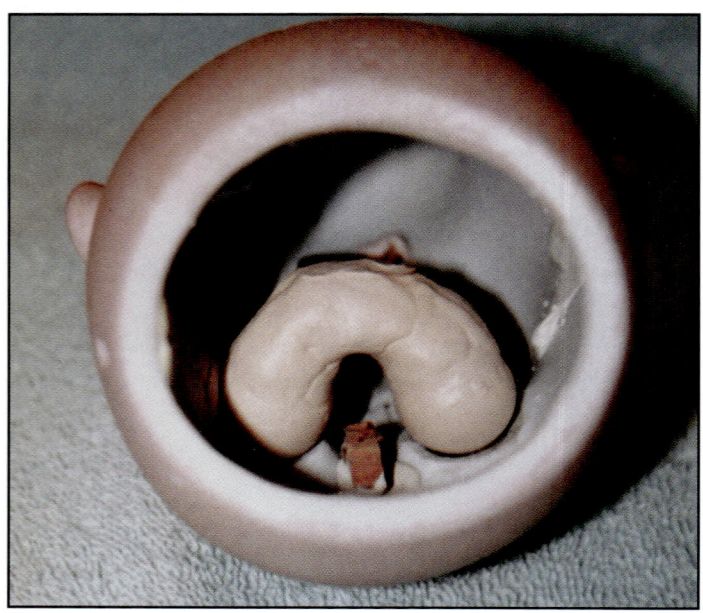

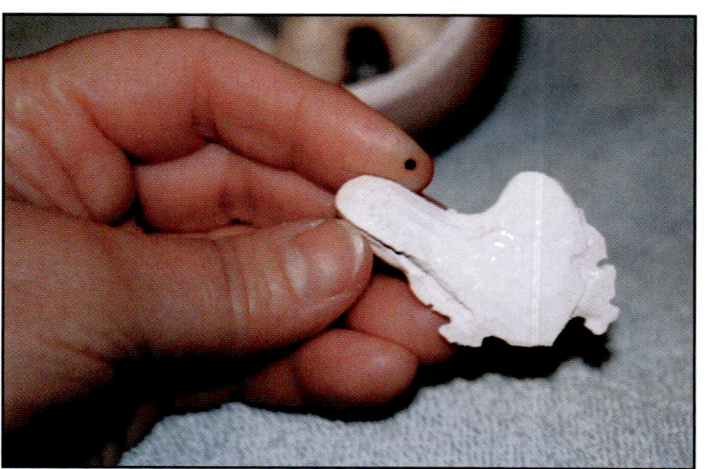

Prop the head with a portion of rolled toweling at each side so that it stays level and allow several hours of drying time for the glued pivot (see below).

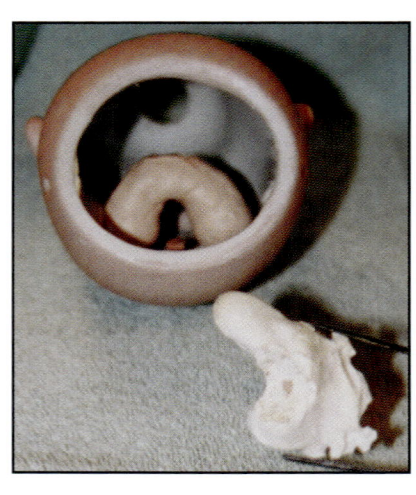

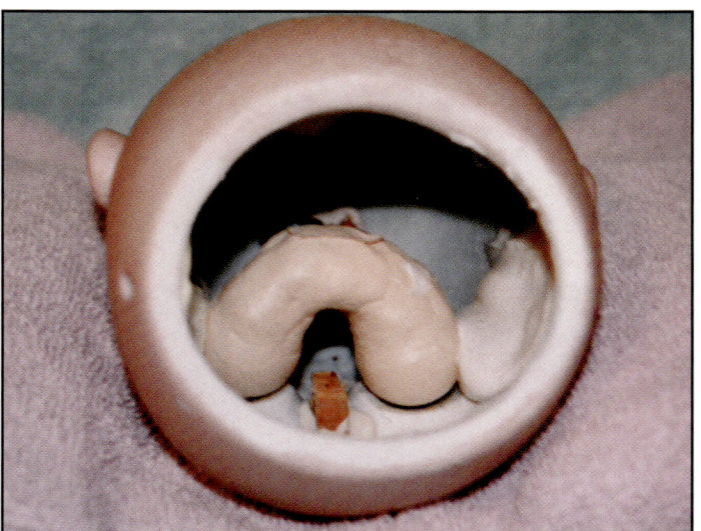

The photo of the finished BBII, with her original eyes, lids repainted and re-set to open and close. Doesn't she look happy?

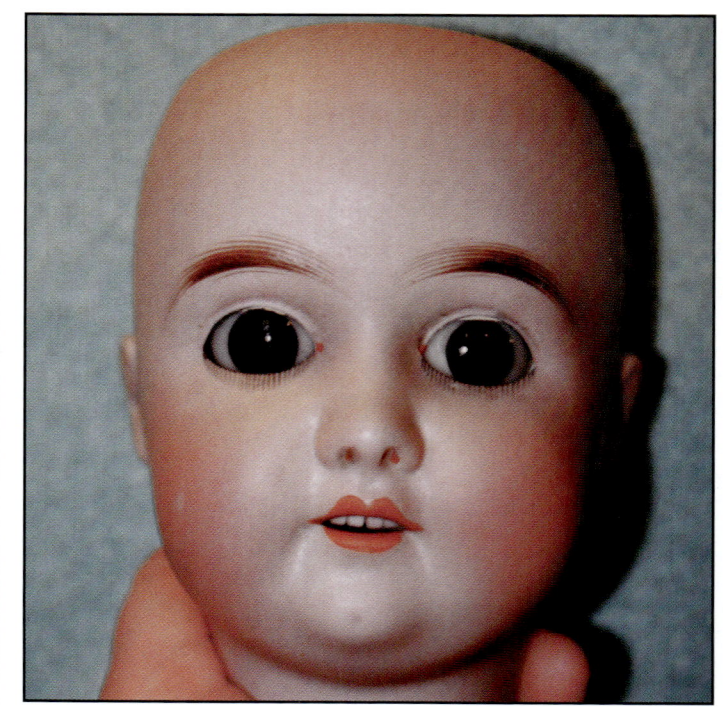

Add a thin layer of Tacky glue to the second eye pivot, which is out of the head and, using large tweezers, place it into the head next to the eye. Adjust it manually and check the fit from the front: holding the glued pivot in place with one finger, rotate the head to assure that the eyes open and close easily (see below and top right).

Original Sleep Eyes Broken, Some Pieces Missing

In this repair situation, the eyes are out of the head; one is broken away from the rocker with two pieces missing.

The pieces are cleaned with isopropyl alcohol. Gluing up the pieces that are present leave a large jagged area of eye missing (see below). Otherwise, the eye alignment is good, and when the eyes are open, only a slight portion of the missing area is visible. What to do?

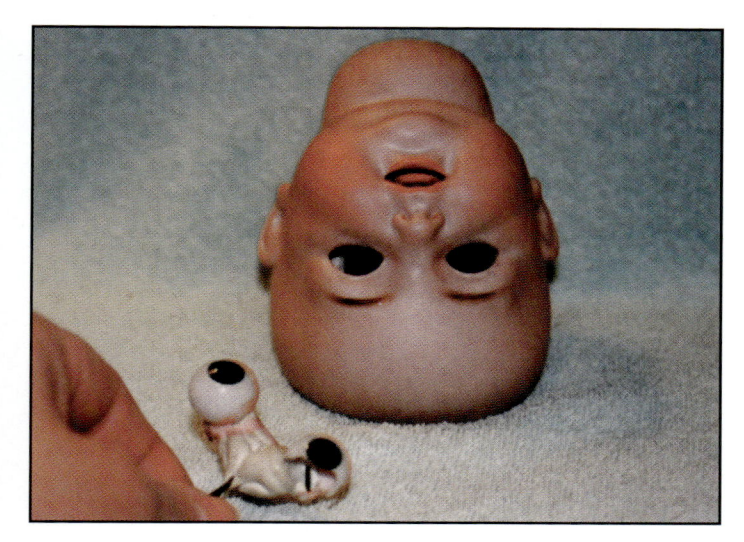

The doll is a significant character baby by Kammer & Reinhardt, mold # 122. In a lesser doll, it might be tempting to replace the eyes with new glass eyes. This more rare doll seems to be worth the extra effort to "save" the original eyes.

Plaster is mixed and applied to the open area in very small amounts. Drying time is allowed between each application. As the repair begins to take shape, a wet paintbrush is used to smooth and round the plaster to the shape of the eyeball.

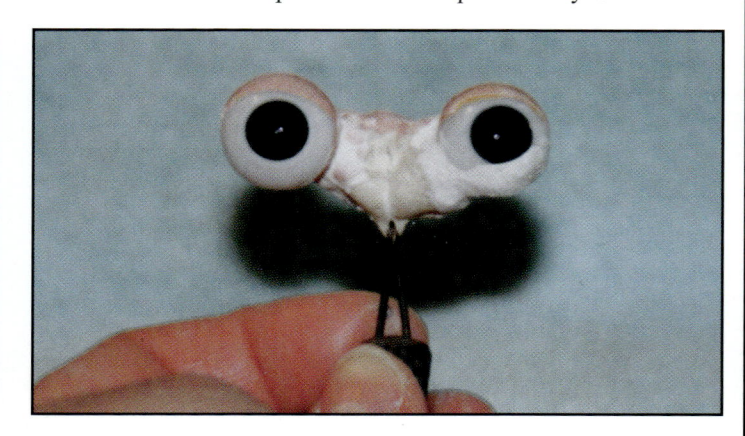

When satisfied with the shape of the repair and after final drying, a light sanding is administered. This gives a final smoothness to the repair and assures that it is tight and will not dislodge.

After a final alcohol cleaning, the eye assembly is inserted into the head and the original plaster pivots installed. (Use the methods previously presented in this section.)

The eye repair is barely noticeable when viewing the doll. It seems the best alternative we could give this unique baby.

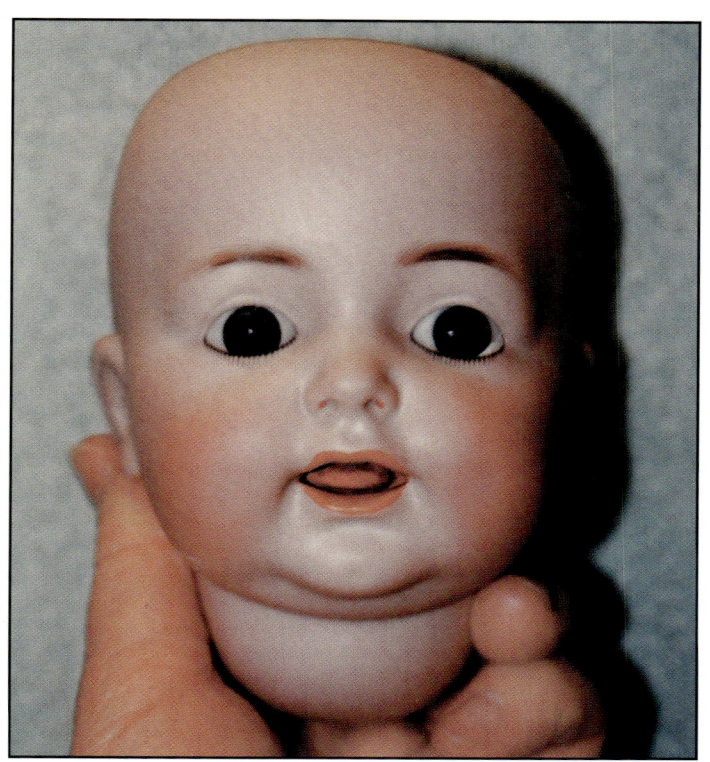

Applying Replacement Eyelashes

While eyelashes, the bigger the better, are the current look in new dolls, not all antique dolls had, or need, eyelashes.

When it is obvious, as in the photo below, that the eyelashes are missing, the application of new eyelashes will improve the appearance of the doll considerably.

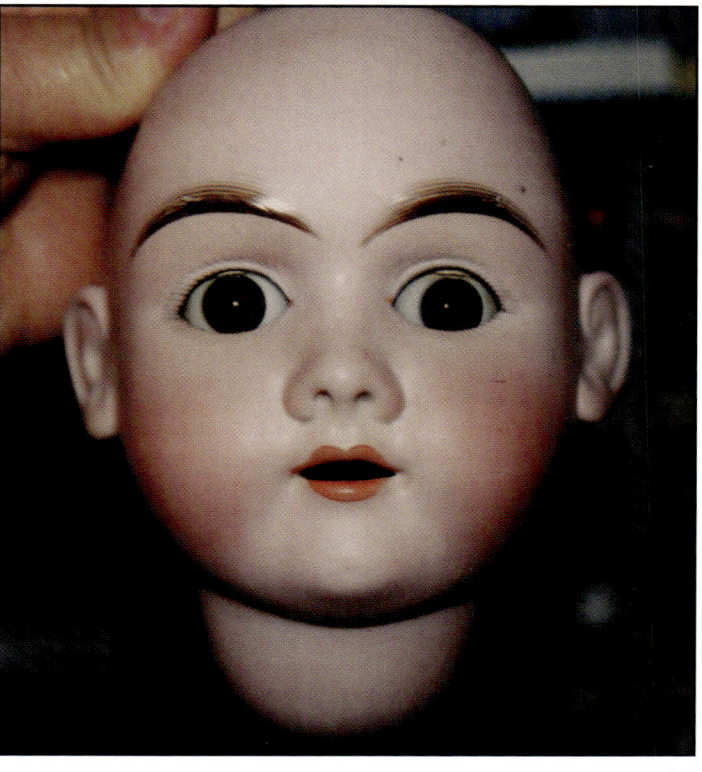

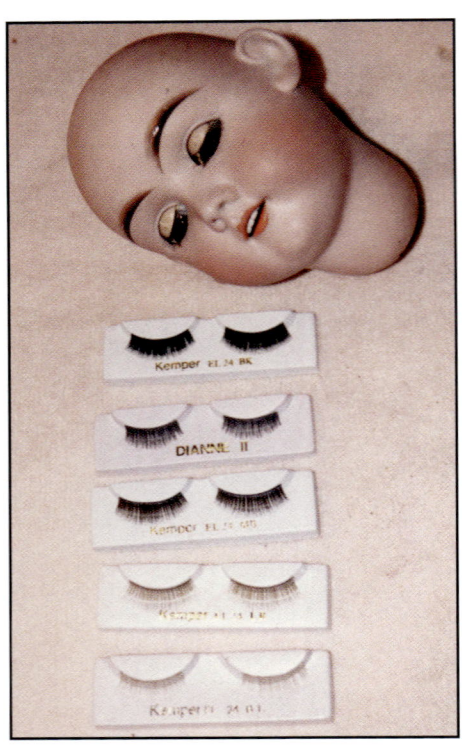

Doll eyelashes come in a variety of sizes, styles, and colors. They are applied, quite simply, by gluing them on with Tacky glue. Clean the eye area with isopropyl alcohol. You may need to trim the lashes to fit the width of the eyelid rim. Cutting the length to shorten the lashes is not advised. Choose a size that is the length you would like.

Have ready a pair of tweezers and your favorite "picky tool", such as a large hatpin or needle. Run a tiny bead of glue along the strip holding the lashes together and set them in place using the tweezers. Then use the hatpin to push them firmly into the paint rim against the eyelid. Finished eyelashes are shown below. This sweet face is an antique Max Handwerck. *Collection of Cheryl Fischer*.

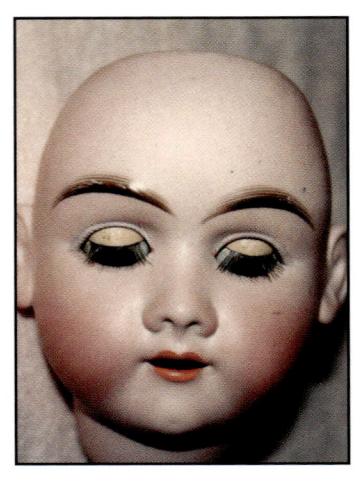

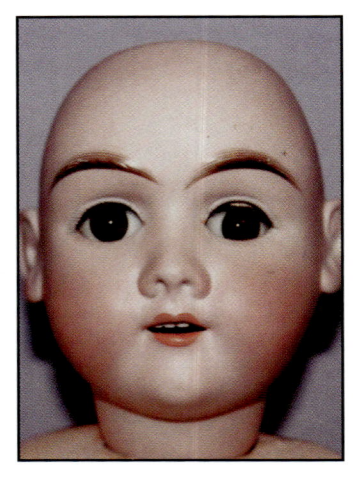

No Eyes Present

When no eyes are present, what are one's choices? Sometimes collectors acquire a doll without eyes, or buy one well-priced, thinking that a set of antique eyes can be acquired and set into the head. This is usually a fruitless attempt. Antique eyes on rockers can surely be had, but it is unlikely they will be the right fit. Each pair of eyes was set on the rocker wire differently to accommodate a specific doll, and it would be more than miraculous to have them fit well in another.

The usual result is that the seeker of old eyes for the old doll will resort to sawing apart the rocker to be able to set the old eyes separately, possibly on a newly fashioned rocker, or fixed. If the eyes are not broken, putting them on a new rocker would allow them to be reset to open and close. In doll terminology, this is called a "sleep eye". If the old eyes are quite fragile, they may still be saved by setting them fixed. This is not a poor result, if the eyes, and irises, are the proper size and of suitable quality. It does keep the repair materials within the period of the doll.

Alas, the repairer may have to resort to using new eyes. In some circles, this is simply not done. Others may rightly feel that using new eyes (provided they are nice quality, properly threaded, hand-blown glass), allows the doll to be enjoyed by further generations and perhaps acquired at a price within the reach of more collectors. Again, it is stressed that this is a reversible repair which can always be easily removed and improved when future conditions might allow.

New Glass Eyes

One excellent choice of glass eye for antique dolls would be Karls Glass Eyes, a German product distributed in the United States by Playhouse. They are beautifully threaded and matched and come in appropriate color choices and a wide range of sizes. Other good sources are also available and all are advertised widely in doll making publications.

If new glass eyes are to be used, the size must be determined. A doll hospital offering good customer service can help with sizing and color selection. If that avenue is unavailable, the eyes can be ordered by catalogue, and these companies usually have a diagram showing how to determine size. Eyes are sold in millimeters, generally even numbers from 6 to 26 mm. It can be helpful to keep a few pairs of inexpensive acrylic eyes on hand to try for size, before ordering the more expensive glass ones.

The new glass eyes are set fixed in the same manner as previously given. Problems with fit and match are virtually eliminated, because newer technologies and customers' expectations have led to a much higher quality product.

Photographs of new and antique dolls, with high-quality new eyes, are shown in the section on new wigs, in Chapter Six, "Wig Care".

Acquiring the correct eye size for your doll is simple when you know how to measure for them. Doll eye measurements are given in millimeters (mm) of diameter. Don't let this metric measuring throw you. You certainly can locate a tape measure or ruler which has metric as well as English measurement. There is no need to try to convert to inch measurement.

There is the need for two measurements to determine eye size: the size of the entire eye and then the size of the iris. Doll eyes are commercially available in sizes from 8mm to 26 mm.

This is quite a range. Most parts catalogues do not usually give the iris measurement, only the measurement of the whole eye. The diagrams for measuring eyes in these catalogs often do not represent the best method.

The simplest thing to do, of course, is to take the doll head to a shop which sells doll eyes. Try some for size, even in acrylic if that is all they stock, and then you will know the size to order in glass.

Tiny New Glass Eyes

Smaller dolls of course require smaller eyes. Eyes of 4 and 6 mm sizes are more difficult to find, and you may feel pressed to use the acrylic ones. Glass eyes in these sizes are available, sold "on a wire" as seen below. This wire could easily be fashioned into a T-frame for the sleep mechanism, but the small size would make this a job for one with some previous eye experience. The techniques are the same, listed below.

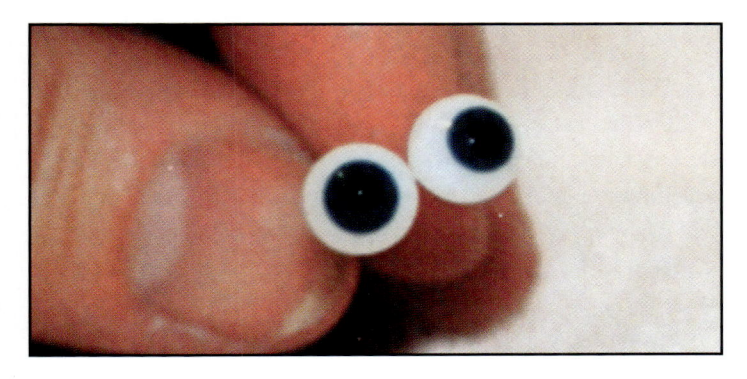

This small bisque is happy she can see dolly things again!

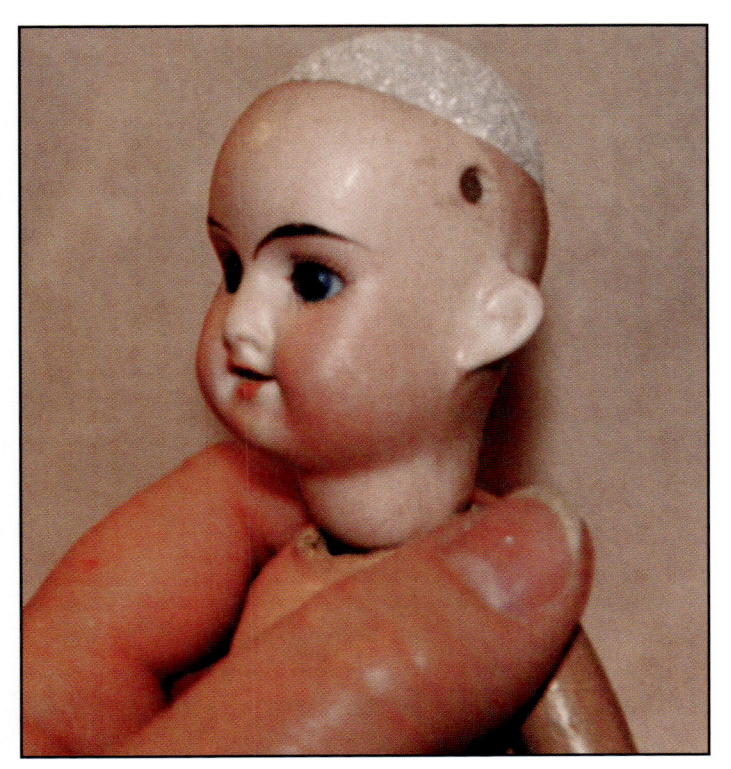

Setting New Eyes to Sleep

Setting new eyes to sleep is in some ways easier than working with old eyes. There are no "inherited" mistakes to correct.

Making a Rocker, or "Eye Frame"

In order for eyes in antique or reproduction bisque heads to "sleep", they must be on an assembly we call a rocker. This is a wire frame shaped like a "T" with a lead weight at the bottom of the T. The eyes are mounted at the two (turned) ends of the top of the T.

Making the eye rocker is quite simple. It's getting it to fit properly that may take a little more effort. Using heavy steel wire and a lead fishing weight, fashion a rocker as follows:

Imagine a "T" inside your doll's head. The vertical stem must measure from between the eyes down to about the chin area. The horizontal stem must measure from the center of one eye to the center of the other eye, plus about 1/2-inch more. Add these numbers and double that amount. This length will allow for some excess, to be on the safe side.

Let's say we are starting with a wire that is 12 inches in length. Thread on a lead fishing weight and fold the wire up around it, wrapping it tightly as you go. You will probably need to use needle-nose pliers to get a tightly-turned spiral. When you have about two to three inches of a spiraled twist, check this "T" against the front of the doll's face for size (see below). The lead weight is at about chin level. The horizontal wires should be crossing right at the doll's eye level. Picture the tip of each of the extended wires being turned so that it goes right into the center of the eye opening.

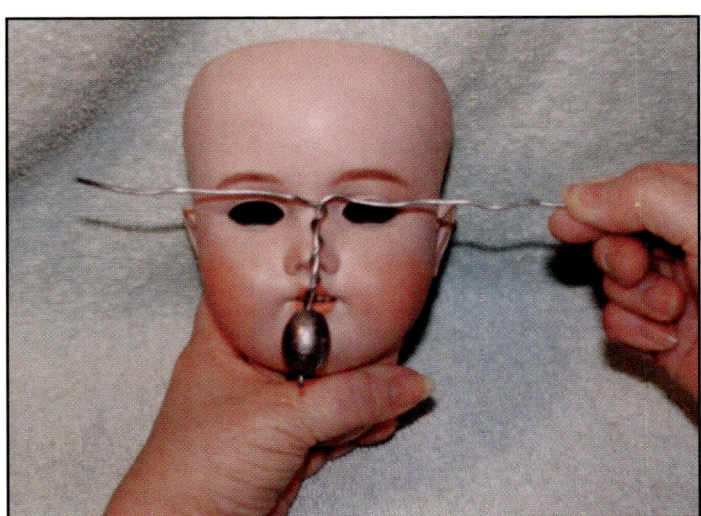

Make any adjustments you need by reshaping your wire "T". If you are really off in dimensions, it will be easier to fashion another "T" than to correct this one.

When you have a "T" which lines up with the chin vertically and the eye openings horizontally, turn the stem ends so that they would enter the eye openings. Cut off any excess so that these turned stem tips are about 3/8 of an inch.

Fix the eyes you have chosen into the socket openings with a small amount of sticky wax on the front of each eye, positioning them so that they match and have a pleasant gaze. The use of an eye holder helps to roughly position the eye in the socket (see below and right).

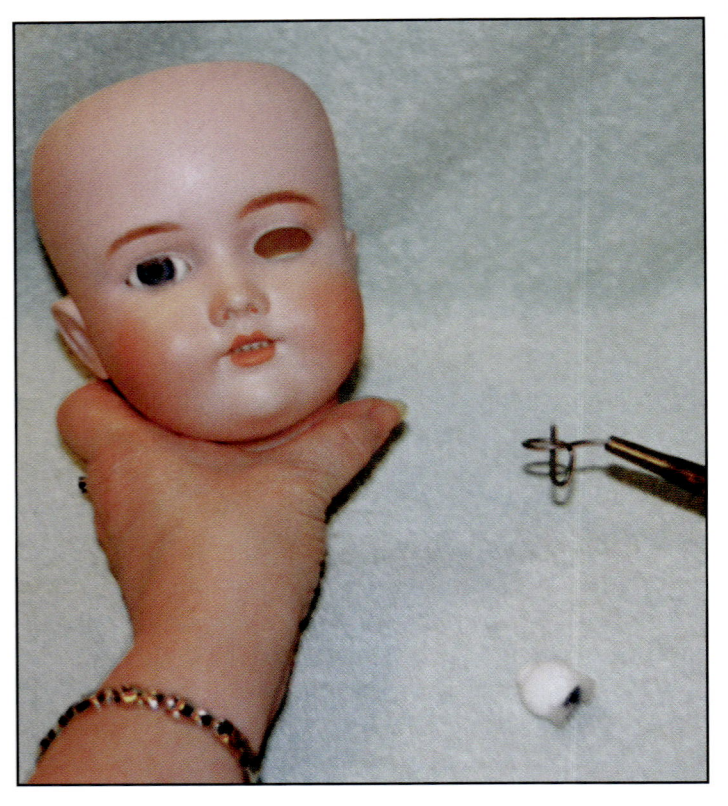

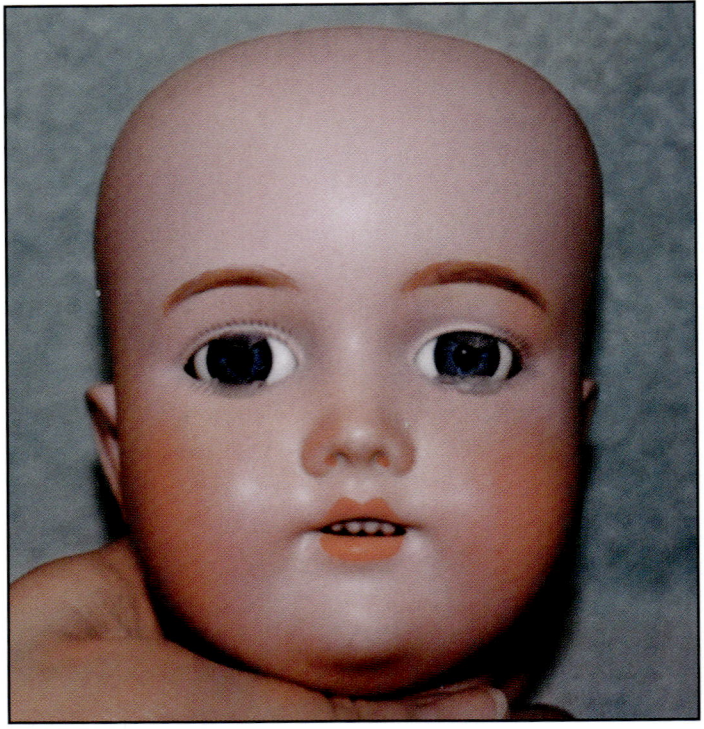

On the inside of the doll's head, position the "T" so that the stem tips go into the stem openings of each eye and the weight rests in the chin. A small piece of styrofoam, firm sponge, or cork will be glued under the weight so that it cannot damage the porcelain. The thickness of this "pad" also helps to align the eyes; it "stops" the eyes from closing too far, which is why it is often referred to as an eye stopper (see below).

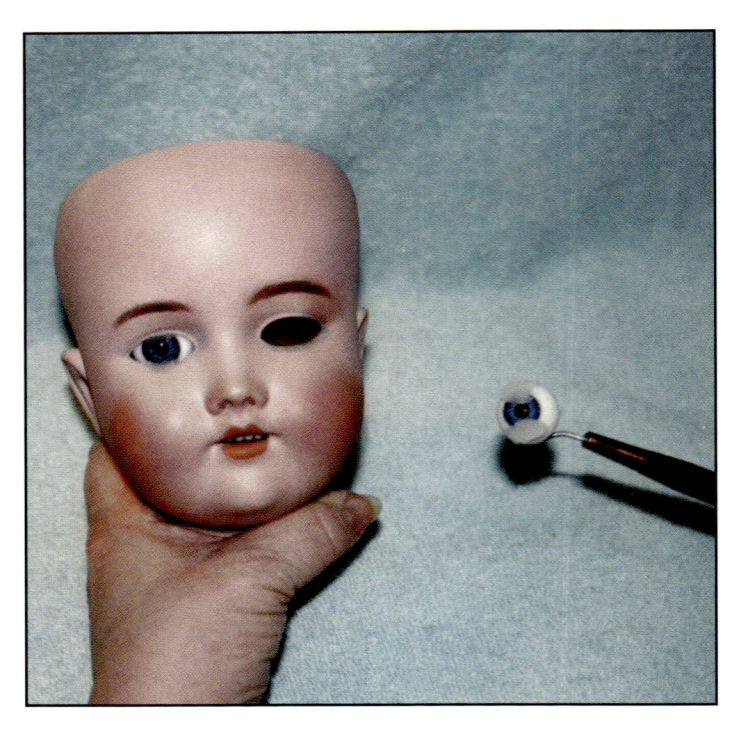

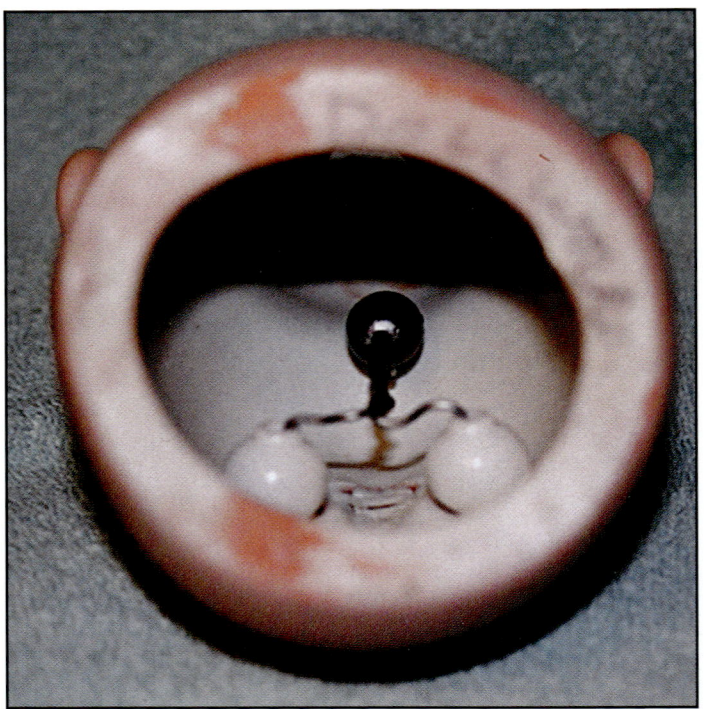

Begin adding putty to the rocker frame between the eyes, a small amount at a time, allowing drying times of several hours between applications (see top photo, next page).

The remainder of the procedure for preparing the rocker is the same as presented for antique sleep eyes in this chapter.

Using a tweezers, lay out hair lashes in a thin but close line on the glued string, adding Tacky to each clump as it is deposited. Arrange the lashes with a clean toothpick as needed. Allow drying time of two to three hours.

Peel the lash/string from the wax paper. Trim with a small scissors as needed.

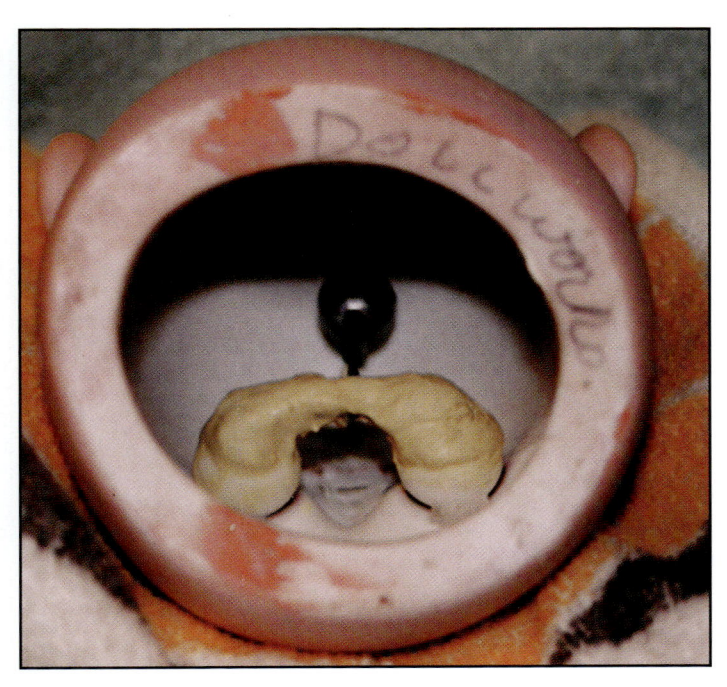

Using tacky glue, glue each eyelash segment to the glass eye at the marked line.

Making Hair Eyelashes for New Glass Sleep Eyes

Before the eyelids are dipped for painting, the opportunity is taken to add eyelashes to these new sleep eyes. On the clean glass eyes, mark the line of the upper eye socket on each eye using a thin black marker. This line will easily wipe off, but is needed to position the eye lashes correctly.

On waxed paper, lay out a length of dark string which has been dipped in tacky glue.

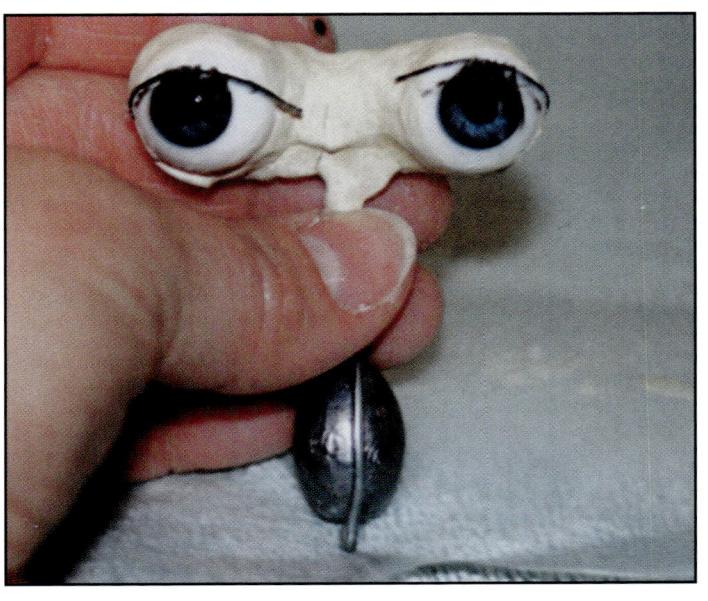

Clean the marker line and the eye in general with a cotton swab dipped in alcohol.

Dip the eyes to paint the eyelids, allowing the paint to just cover the line of the dark string. This not only conceals the string, it also strengthens the lashes adherence to the glass eye. Allow drying time as usual.

Installation of a New Glass Eye Rocker

This involves "pouring" the plaster pivots since there are none remaining from the old set of eyes. Even if the pivots were available, they would not fit, since they were made for a specific eye rocker.

This is a task feared and avoided by most doll restorers, which is why all those mis-fitting eyes are seen! It really is not that difficult, but takes a little practice to master. The worst thing that will happen is that the rocker won't move because too much plaster was used and the work will need to be done over. (Which will give you some practice!)

Position the eye assembly in the head and tack into place with sticky wax on the front of the eyes.

Mix a small amount of plaster with water to a thick, soupy consistency.

Coat the side of each glass eye with a thin layer of liquid soap, Tincture of Green Soap, if available, is the preferred product, but any liquid soap will be suitable. This layer of soap prevents the plaster from sticking to the glass as it dries.

Stabilize the position of the head by rolling up a hand towel at each side of the head.

Using a baby spoon, "lay" a spoonful of plaster at the side of the eye, allowing it to cup about the eye slightly, moving the spoon up the side of the doll head, toward the ear (see below). It takes a bit of experience to master this technique; once you have done it a few times, you will know just how much plaster to use, and the task will become easy for you.

Allow overnight drying time. Remove the sticky wax from the eye fronts, then wiggle, jiggle, and gently coax the eyes into slowly...slowly working open and then closed (see below).

If you simply cannot move the rocker at all, soak with water and remove one of the plaster pivots. Hold it in place with a finger, move it ever so slightly outward, and check again by rotating the eyes, whether or not they will open and close. If this procedure cures the problem, the eyes were just a little tight. Glue the pivot in place and allow to dry.

If this procedure does not cure the problem, the eye pivots are surrounding the eye too much and need to be re-poured, smaller, especially toward the top of the eye.

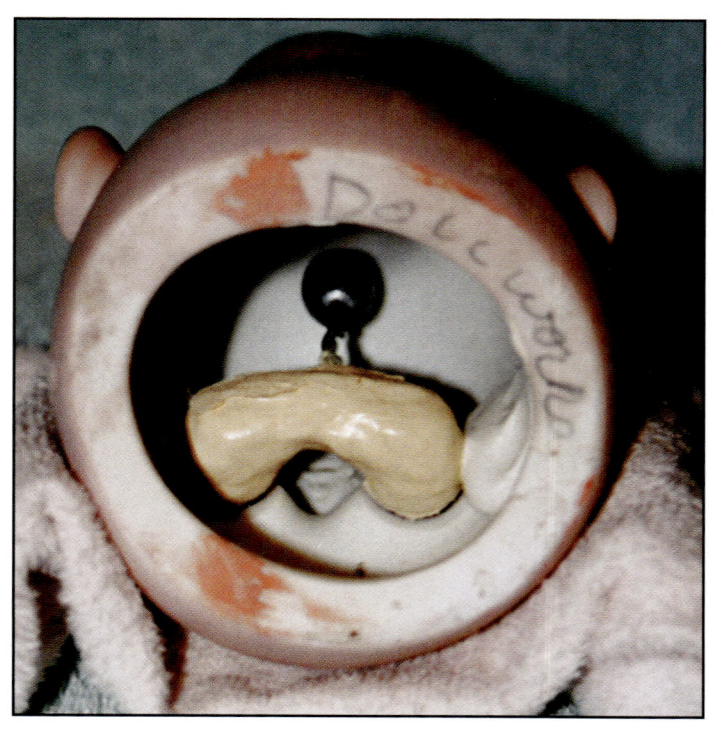

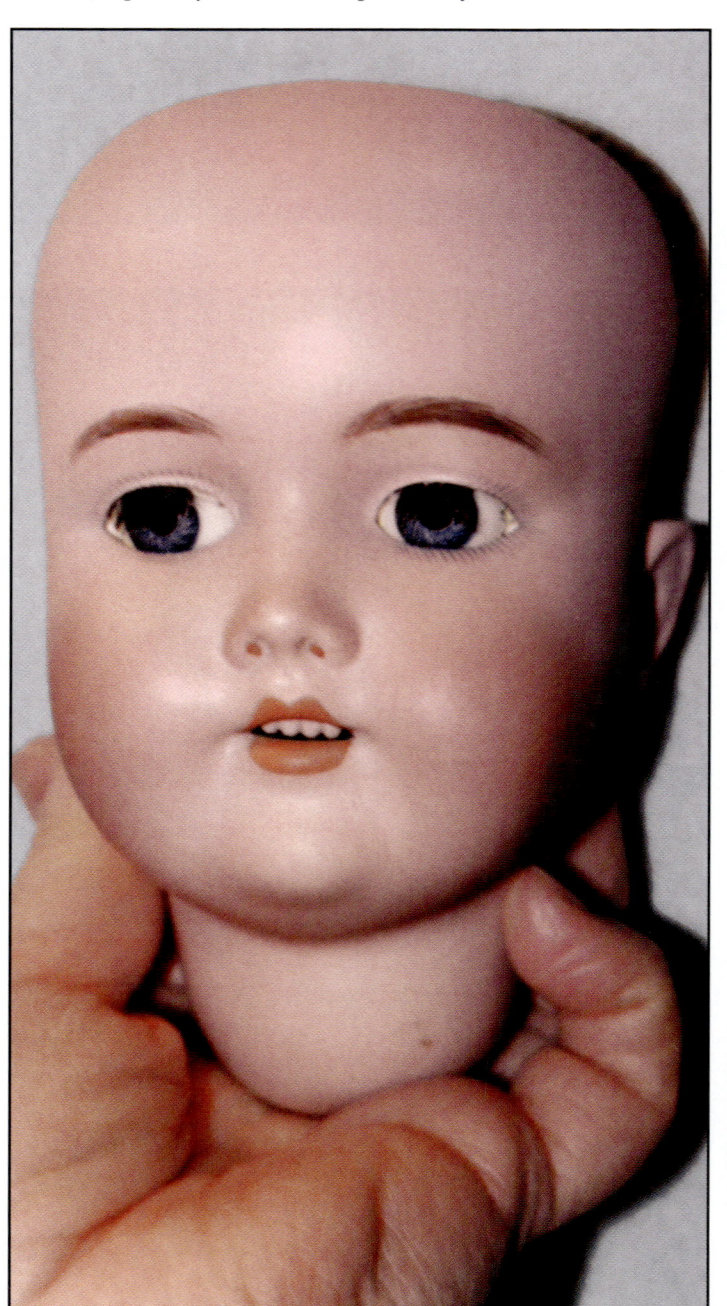

Wig Care

Types of Doll Heads

Many antique dolls which are wigged are open-crown bisques. The doll head is designed with an open access which slants down and backward from the forehead toward the nape of the neck. This design was intended to give adequate access to install the sleep eyes. This open area is first covered with a concave rounded shape, usually made of cardboard, sometimes poured or plaster or a similar substance. This "pate" is glued on prior to gluing on the wig. French antique bisques are noted for pates shaped of cork. The French dolls usually require a deeper pate, as the slant of the open crown is usually much lower than the German dolls.

Some antique bisque dolls which are wigged will be closed crown, that is, the head is fully shaped of porcelain, similar to the infants. This eliminates the need for a pate, as the wig can be glued directly to the head.

Types of Wigs

When evaluating antique dolls relative to the appropriateness of the wig, several considerations are in order. The wig may be the original, may be a handmade replacement wig, or may be a commercial replacement wig. The original wigs were made of natural hair, either human or goat (mohair), or sometimes horsehair. It is important to learn to distinguish these materials from the synthetic wigs, widely available since the 1940s.

With experience, the restorer will "know" whether or not the wig is original. The methods of manufacture used will become familiar. Obviously, if the wig is synthetic, it is not original to an 1890s doll. The synthetic wigs came on the market after the ending of World War II (around 1948), when they also became common on the hard plastic dolls being made then.

Wig Shampooing

Basic supplies include:
Shampoo
Conditioner
Old towels
Metal dog comb, such as Hartz
Small sharp scissors
Various sizes of permanent wave rods
Permanent "end" papers
Hairnets
Various sizes of hair clips for holding excess hair out of the way and positioning some curls.

White plastic wash basin (dish pans or kitchen bowls are suitable). The white color is important so it will be easy to see when the wig is getting really clean. The need for white basins is more believable once it is tried.

It's helpful to have several wig stands, made with styrofoam balls of different sizes mounted on a dowel rod in a wood base. Cover the styrofoam with plastic wrap to prevent the wig hair from getting caught in the styrofoam surface.

It should be mentioned here that for those who believe in the reward of heaven, special status there has been reserved for persons who shampoo and set doll hair. Of course, there is also the earthly reward of performing frequent miraculous feats of coiffure. Often, dirty and disheveled doll hair responds very well to this treatment, and the one wielding the soap and comb will be a hero or heroine to the members of his or her acquaintance. So, take a "before" picture and then march that dirty dolly to the sink at once!

A Note of Caution

The instructions in this volume pertain only to wigs on antique bisque heads. Removal of wigs from composition/hard plastic heads takes quite a different approach, and will be thoroughly discussed in that volume. Composition heads can never be soaked in water. In addition, the wigs on these and the hard plastic dolls may not be wefted wigs, and might come apart if washed. The care of these kinds of wigs will be thoroughly covered in another volume.

Unless the doll has been remarkably well stored, the wig is undoubtedly due for a shampoo. Imagine 80 or 90 years of dust and dirt are contained in it. First it must be removed from the head. For several reasons, it should be soaked off rather than merely pulled off, unless it is very loose and comes away easily.

Removing the Wig from the Doll Head

The wigs original to these dolls were sewn to light-weight cotton netting, which is often not as strong as the glue which may be holding it. If pulled

Chapter 6

away it may be badly torn. It is also possible to cause "wig pulls" to the bisque; these are white spots where the color has actually been pulled away with the glue. If the head is off the doll, it is quite easy to soak it in a small padded basin, in about an inch of warm water. It is important not to have the water too deep, as it would soak loose the plaster curves which hold in the doll's eyes. If the head is attached to the body, improvise a way to soak it. One favorite is to gently tie the doll's arms to the sides and put her upside-down in a stand. It may appear she is being punished for some unknown doll naughtiness!

While dolly is indisposed, there is time to prepare a drying stand for the wig. Acquire a styrofoam ball the same size as the doll's head (or slightly smaller) Attach it to the upright of a doll stand, or a piece of dowel rod glued into a wood base. It is helpful to cover it with a piece of kitchen plastic wrap so the wig doesn't get snagged on the styrofoam. This apparatus will insure that the wig base doesn't shrink as the wig dries.

If the pate comes away from the head with the wig, leave it to soak with the wig during the cleaning. It may disintegrate, but can be easily replaced. Trying to remove the pate from the wig while dry may cause damage to the wig base. While it's wet, handle the wig very gently.

Shampooing a Doll Wig

Once the wig is removed from the doll's head, determine whether it is natural fiber (horsehair, human hair or mohair were originally used), or whether it is a synthetic replacement wig from the 1950s or later. Soak the wig in a small basin in warm water, to which is added a small amount of shampoo, assuming the wig is natural hair. If the wig is synthetic, such as modacrylic, shampoo is not recommended as it may contain conditioners which relax the fiber too much to take a set. Plastic wigs should be washed in whatever mild dish detergent is on hand.

It is always amazing how dirty some doll wigs actually are, even when they don't appear so. The wig may take several changes of water and several hours or days to come clean. Remember, some of this dirt has been there for scores of years; give it some time to soak loose. Gentle squeezing in each new water will help move things along. Also, there may be a quantity of old glue to loosen, as well as perhaps the pate. Let the pate soak loose rather than trying to pull it away from the wig, lest you tear the delicate net base.

When the wig seems clean, rinse, again gently, in several changes of water. Let an hour or so lapse between the last two or three rinses to allow thorough diffusion of the water.

The Use of Conditioners

Conditioning products of the type that would be used on your own hair are not recommended for synthetic wigs. If a synthetic fiber is wiry and very difficult to comb out, a conditioner may be used in this step, but then rinsed away before setting. Left on the wig, conditioner renders synthetic fibers limp and the set will not hold well at all. Also, it seems to leave a "greasy" feel to the fiber and, of course, since it will not be shampooed frequently as would a person's hair, it attracts dust.

Conditioners are well put to use on real hair wigs, whether horsehair, human or goat (mohair). Horsehair in particular really benefits from its application. The conditioner seems to be absorbed by the real hair fiber, leaving it soft and receptive to a long-lasting set.

In any case, whether a conditioner is used or not, rinse the doll's hair well with (hopefully) soft water. Squeeze as much water out as possible, then roll in old terry toweling and again squeeze as dry as possible. Handle gently as you would a fine garment, squeezing and blotting rather than wringing.

Drying the Doll Wig

Next, gently shape the wet wig on the drying stand prepared for it, with one or two paper towels under it to absorb the remaining water. It may still drip, so place it, stand and all, in the (empty) basin in which you washed it.

If a drying stand cannot be made available, stuff the wig gently with paper toweling, as close to its proper size and shape as possible. Lay it out on toweling to dry. Be particularly careful in handling old wigs when wet. After a thorough soaking, the glue that was used (and was possibly holding the backing together) should all be washed away. This leaves the net backing quite limp and loose in some cases.

Short Baby Wig Shampoo and Comb-Out

The photo below shows a short baby wig of mohair, soaking in a white plastic basin. If you look closely, you can see deposits of soil in the bottom of the pan.

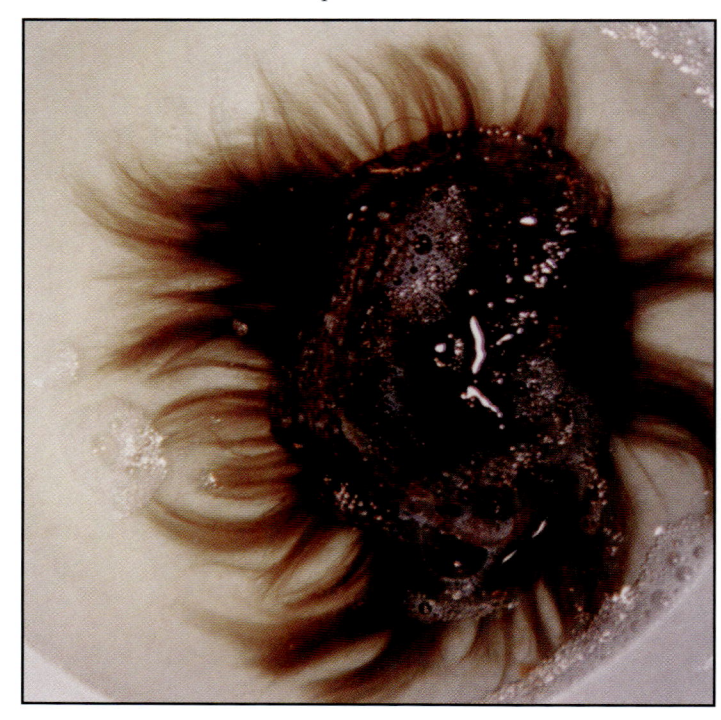

Change the water/shampoo as often as needed, until the water remains clean.

Rinse several times, using a small amount of conditioner in the next to the last rinse.

After the final rinse, place the wig on a wig stand and allow to dry (see below).

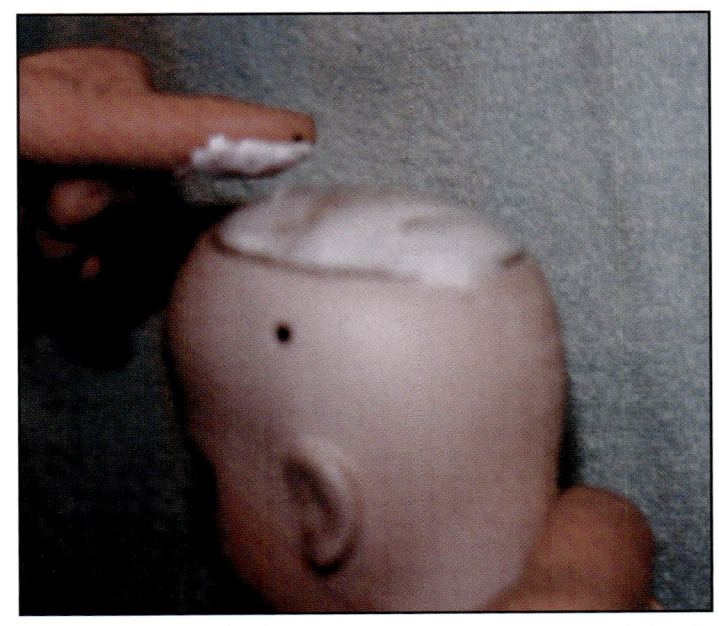

Roll the wig rim back as though you want to turn it inside out. Position it correctly for front/back.

How to Glue a Wig On a Doll

Glue a pate on the open crown head and allow to dry for several minutes. Have a wet paper towel at hand. Then apply Tacky glue to the pate only.

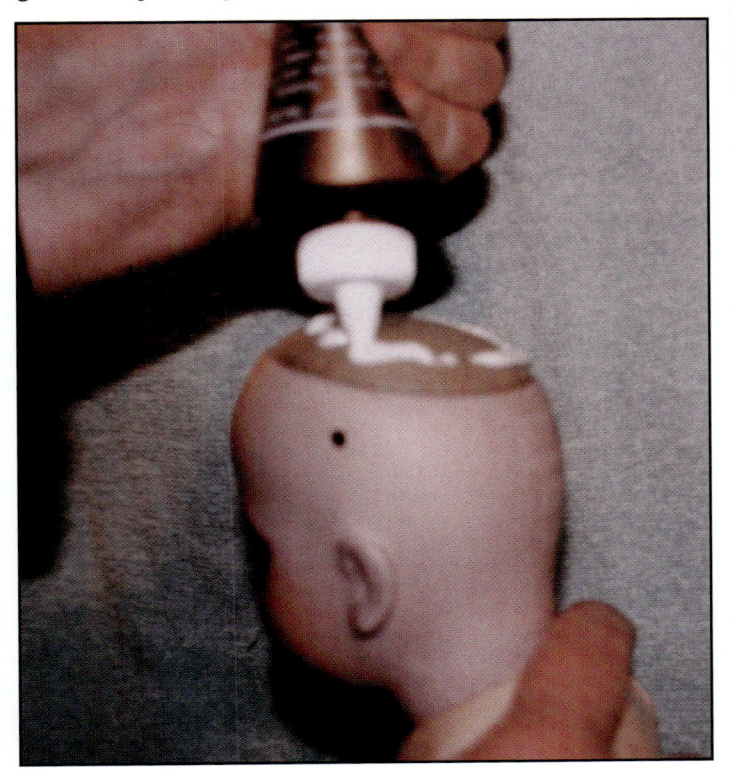

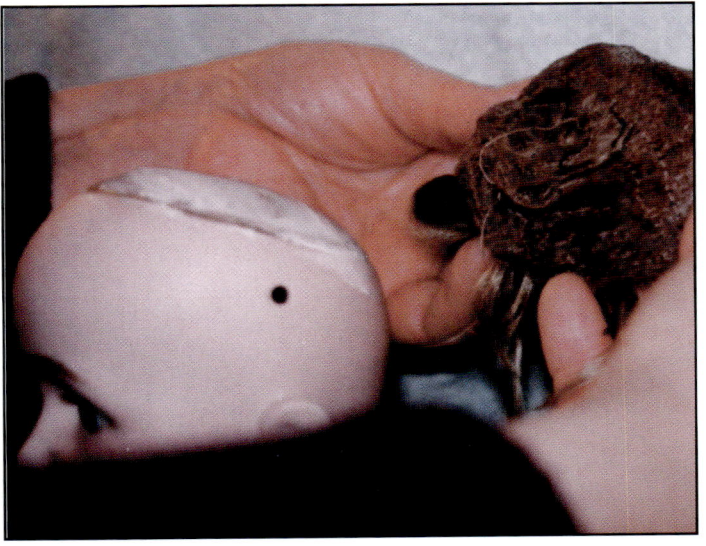

Roll the wig onto the doll's head.

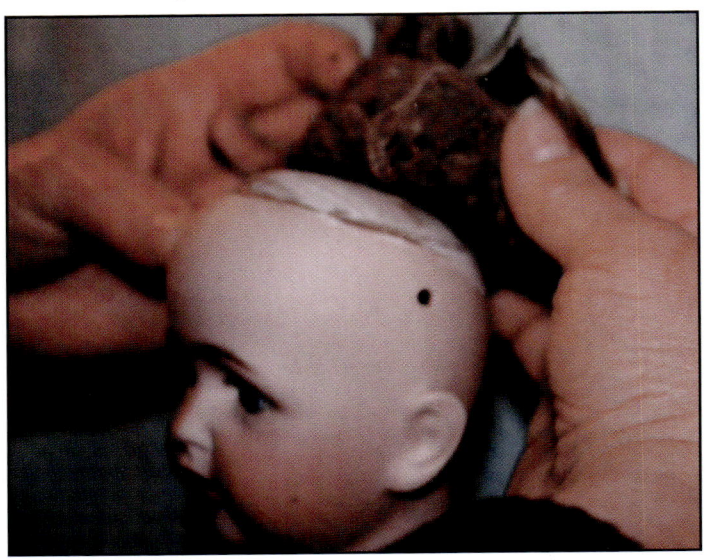

Spread out the Tacky glue with a finger, just enough beyond the pate that it will be covered by the wig base. A thin layer of glue is all that is wanted; wipe away any excess and of course clean off your finger with the wet paper towel.

Before the glue sets up, position the wig correctly, bringing the edge of the wig base (the netting) down as far as it will come and making sure that no hair is caught up under the wig.

Allow a few minutes for the glue to set up. Ever so gently, pick and comb lightly at the mohair, using a little spritz of water as necessary to make the task easier.

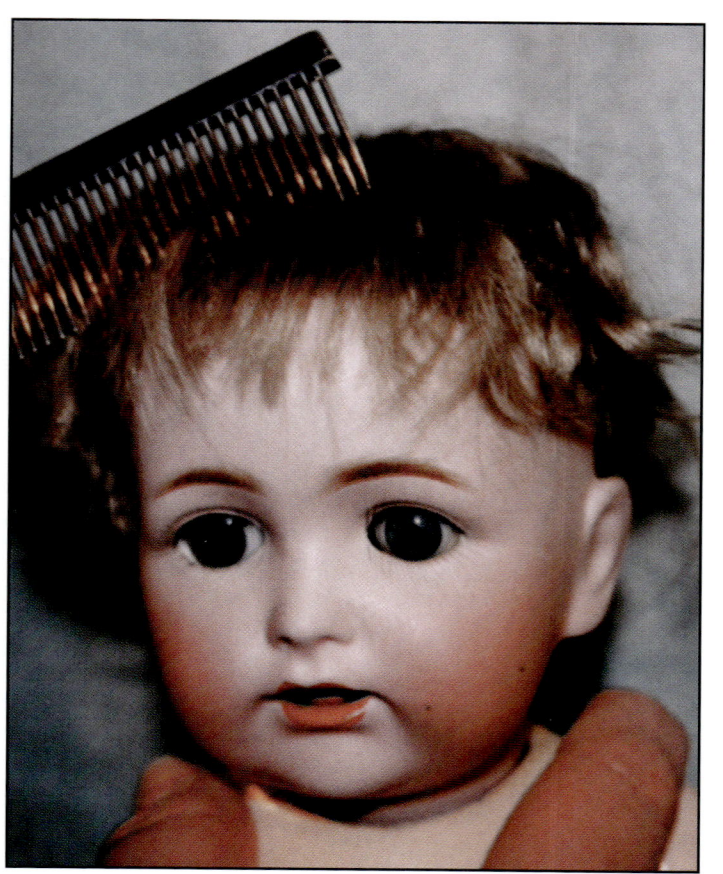

Place a loose hairnet over the hair to "settle" any loose ends, using a little more water mist if needed. Allow to dry.

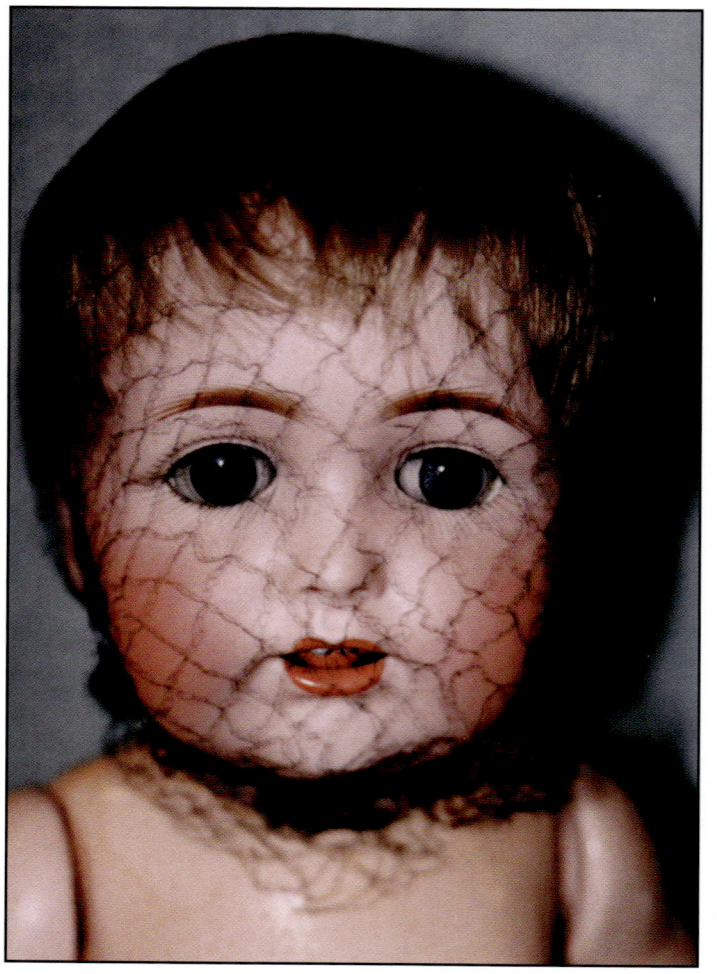

Wig Mending and Repair

After a wig has been shampooed and is dry, check the inside backing carefully; some areas may need mending, or some of the wefts may need to be re-stitched to the backing. Use thread close to the color of the hair, and with darning techniques, reinforce any weak areas.

Be a brave little soldier here...some wigs will look quite hopeless at this stage, quite like something the cat (or dog) dragged in, wet and wild. Take courage. Miracles occur with time and patience in the wig setting business! If the backing has suffered dry-rot and has somewhat "disappeared" during the wash, it can be "darned" back, or even replaced with cheesecloth or netting. Study the way in which the wig was originally sewn together and duplicate the rows as well as possible. Remember, these rows will be a guide in setting the hair.

Wefted hair can also be added where needed, if the proper type and color can be acquired. Mohair wefts are much more available than human hair, and in a good variety of colors. Some will use personal sources of human hair; this is ill-advised for two reasons. One reason would be the obvious sanitation considerations. Secondly, results are usually marginal at best, due to poor condition of the hair.

Reattachment of the Wig

Usually it is best to glue the wig back on the doll's head before setting the hair. Obviously it is easier to arrange the curls about the face more suitably. Use a "tacky" glue, as little as possible. When the wig is clean and dry, do a trial placement (without any glue) of the wig on the head first to catch any possible errors of placement or problems with fit. It's miserable to find a small hole in a wig after glue has been applied to it. This type of "goof" is one that is usually remembered well and never repeated if at all possible. Also, it may take a little adjusting of the wig to figure out how it should be placed on the head, as it is still probably quite disheveled after being shampooed.

When all looks well with fit and repair condition, make a mental note of how far down the wig comes on the doll. A light pencil or chalk mark is allowable to mark where the edge of the wig cap will be. Then place a light film of "tacky" glue on the doll's head, down just far enough to glue down the wig cap. Glue in the hair itself is to be avoided; if this happens, blot away with wet paper toweling, a task that will take some time.

Steps in Applying a Wig
- First, try wig on doll head without glue to check fit.
- Then, wet a clean paper towel!
- Apply thin layer of "tacky" glue to doll head.
- Clean hands with wet towel.
- Turn wig "inside-out" in a rolling motion, turning all the hair into itself, away from the wig base.
- Keep in mind the positions of front and back of wig.
- Roll the wig onto doll head, keeping hair out of glue.
- Adjust position of wig, i.e., how low on forehead, is there a part?
- Apply a couple of rubber bands to make good contact with the glue and let dry for 20 minutes or so before starting to do the set.

Choosing the Hair Style

The doll may have come to you with the hair so disheveled that you really have no idea of what the original style was. A little inspection, looking for part lines, will usually settle the question. Most original sets in antique bisque child dolls had some type of part, either center or side, with hair in long "sausage" curls.

This hair style is set with the curls in a vertical pattern, rolled toward (from the center back of the head) the face of the doll. This is the style on which we will elaborate, as it is most commonly found. A few of these darlings had braided hair, either in "pig-tail" fashion or coiled about or around the head. This requires substantially longer hair, and is not often seen in the mohair wigs. Usually these are found, original, in horse or human hair, as these latter fibers were available in much longer lengths. One tip that the wig was braided is a stitched part down the center from forehead to nape of the neck.

Antique bisque baby dolls of course had simpler hairstyles. Many of the character baby dolls had so little length that only a simple comb-out is necessary, perhaps netting it before using a light spritz of water to "settle" it to the head.

Such luscious babies as the Kestner "*Hilda*" had a bit more hair, and length, and were done in basically one row of large curls about the bottom of the wig.

Wig Setting Techniques

For wig setting, the following "tools" are needed:
Cape or plastic wrap to protect the doll body
Pet (dog) comb having smooth round metal teeth
Home permanent rods
Permanent "end" papers
Spray Bottle for water mist
A few hair clips/hair pins
Hairnet
Lots of patience!

Identifying Wig Fibers

Since this discussion concerns antique dolls, it is assumed that the wig is of a natural fiber, namely human, goat (mohair) or horsehair.

Often, we encounter antique dolls which have had a wig replacement, say in the 1940s or 50s, which to the owner looks quite old. They may believe it to be original. It takes a little experience to always recognize an older synthetic wig, as some of the early Sarans mimic mohair quite well to the novice observer. It also does not help the novice restorer that in recent years, "synthetic" mohair wigs have been touted on the marketplace. Simple to know that when the tag says "synthetic mohair", it is synthetic!

Having once been fooled in this situation, I developed, early on, a sharper eye. Experience is such a wonderful teacher, there is no other lesson quite like it. I entertain this discussion here because, if you are trying to set a wig which you believe to be a real (animal or human) fiber and it is synthetic, you will have quite a time of doing the job, and may get very minimal or no results (that means lasting curls!!).

So, do make this determination first, or at least be aware that it's possible to mis-identify these fibers. It is very difficult to explain how to tell the difference, just as it is very difficult to tell someone the difference between new and old porcelain. "You just know by looking at it" is not very helpful advice.

One possible test to distinguish natural hair fibers from synthetic, if you are at liberty to snatch a few strands, is to burn them. Both synthetic and natural will smell quite awful, but the synthetic will leave more of a melted, charred mass, while the natural fibers will leave more of an ash.

Mohair, human hair and horsehair will all handle quite similarly and give good setting results. Mohair curls will last quite a few years and only tighten with humidity. Human and horsehair also take quite good sets, with the familiar disadvantage of drooping with humidity. Even so, sets usually last ten or more years, probably enough for most of us to be willing to undertake this project.

Hair Setting Procedures

You will need to choose a curler size which would be smaller than those appropriate for humans. Home permanent rods come in sizes that are perfect for doll hair, because they are of much smaller diameters. The use of end papers is a must, first to get a good roll-up and then to remove the rods more easily.

Horsehair wigs tend to set quite tightly, and because of this, use of a slightly larger diameter curler is recommended. Go up one of two sizes from what you would normally use for human hair or mohair.

Setting an Antique Mohair Wig, Antique Child Doll

We have graduated now, from handling the simple short baby wig in the previous repair session to more of a challenge. We have carefully removed, mended, shampooed, conditioned, and allowed to dry the original mohair wig on this large Simon and Halbig beauty. The wig is somewhat fragile and not uniformly full all around. It will still take a set, and style beautifully to be enjoyed for another eighty or ninety years.

Clean, dry wig, front and back views (see below). Begin in the back of the doll's head. Section off what you consider to be the bottom row of curls. Don't be concerned at this point if it is quite tangled, just separate it, using a smooth tool like a very large sewing needle, crochet hook, or one or two teeth of your metal comb, whatever works well for you.

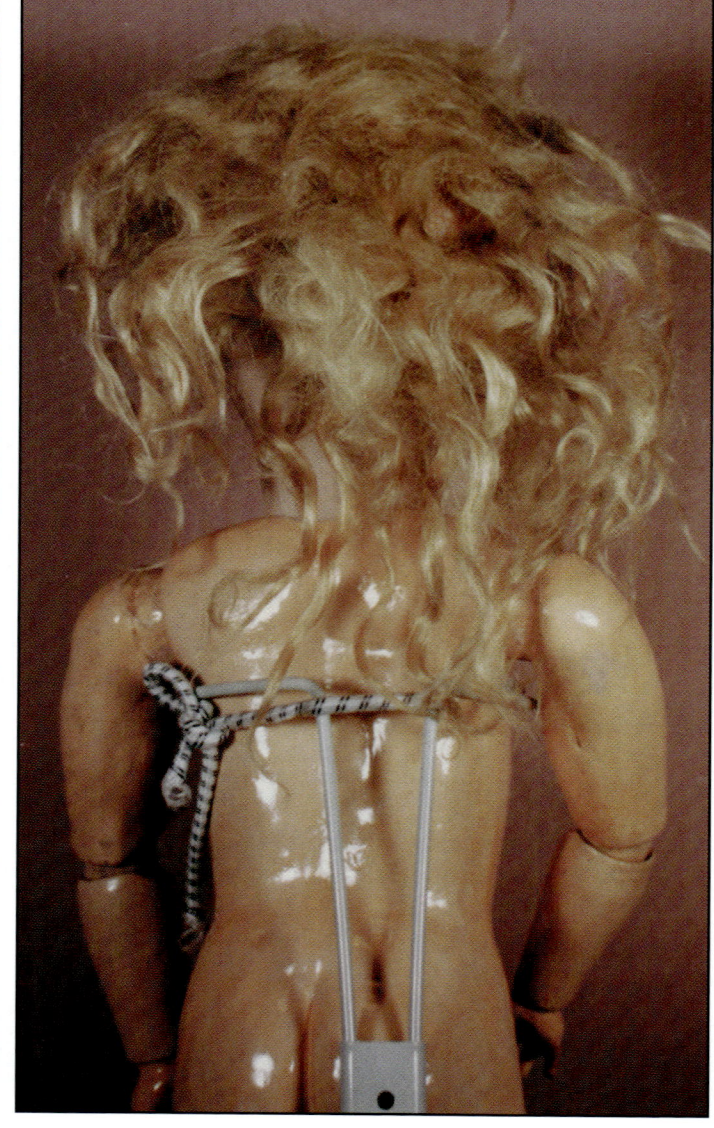

Using large hair clips, clip the rest of the doll's hair out of the way. Drape the doll to protect the body from moisture, with a plastic cape (a large trash bag will serve for this function). Beginning at the center back of this bottom row, separate the first section, holding the rest of the hair away with clips, if needed. Working gently, comb out all tangles. This is the most important step in achieving a smooth and lasting curl. Don't be too concerned if you lose a small amount of hair in doing this and, if the section is quite uneven at the bottom, a slight trim is acceptable. Of course, we would usually say that we don't cut original wigs, but minor evening of the rows may be necessary. When the strand is combed tangle-free, raise the section of hair a bit as you will be doing a vertical curl; fold an end paper about it at the end of the hair and spritz with water. Proceed with the roll-up, rolling toward the doll's face, either right or left on this first one, stopping where the curl can lay easily against the doll's neck. This is curl #1.

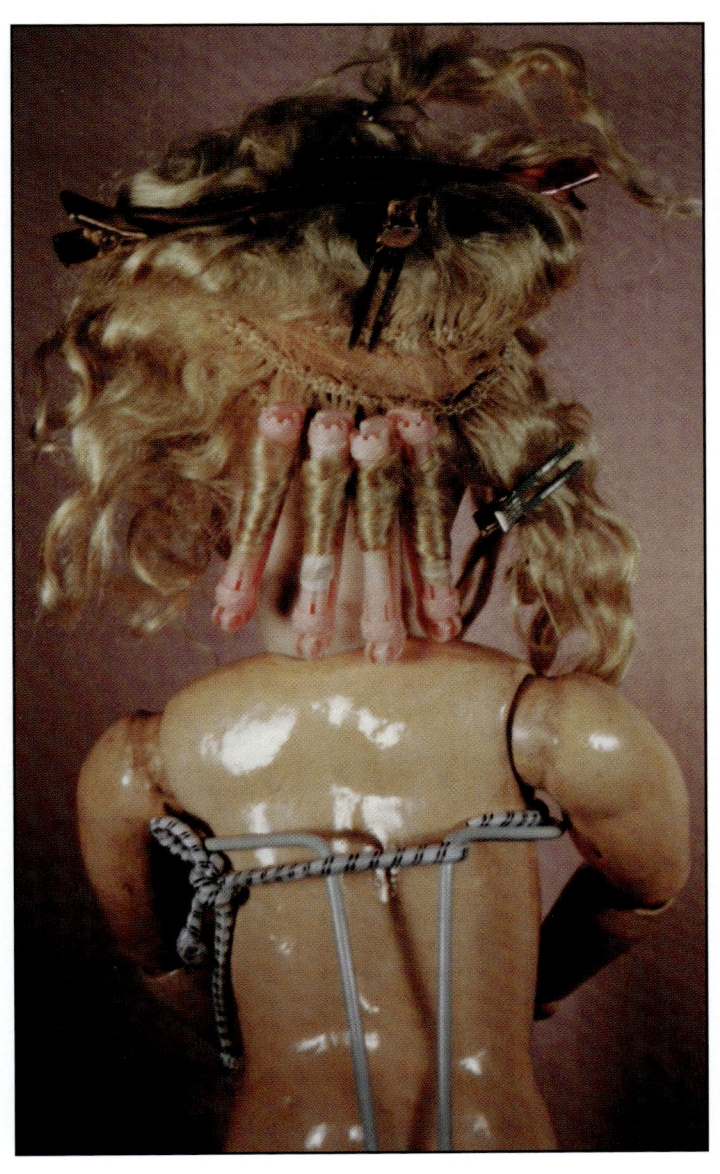

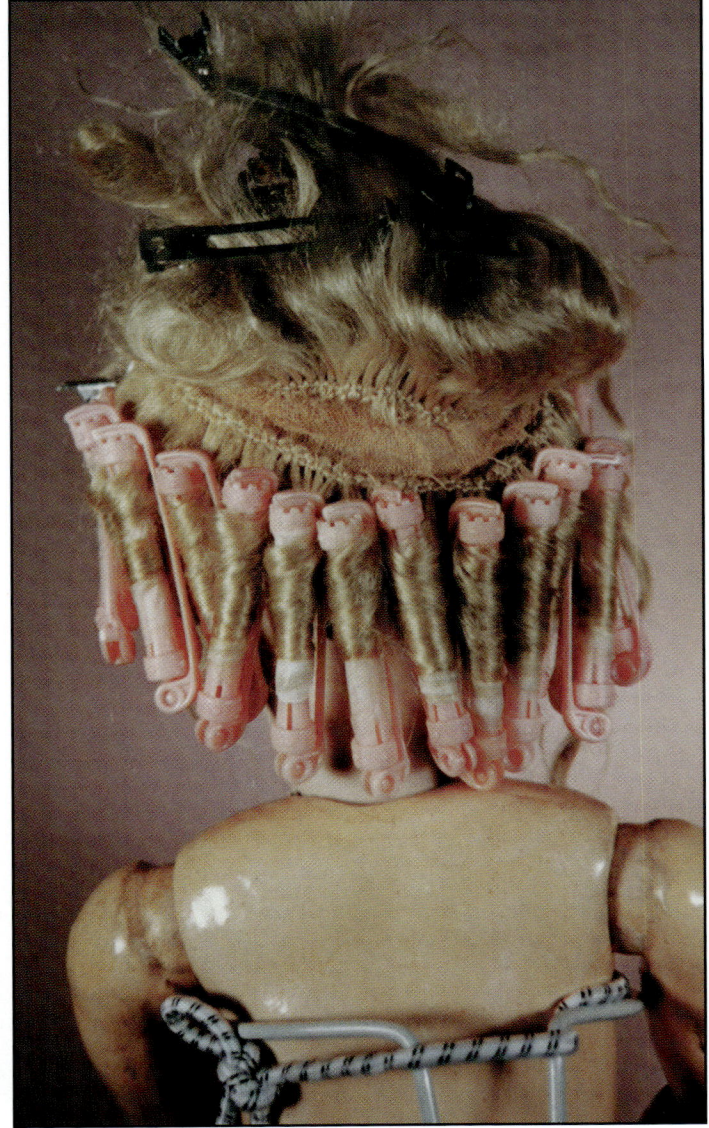

95

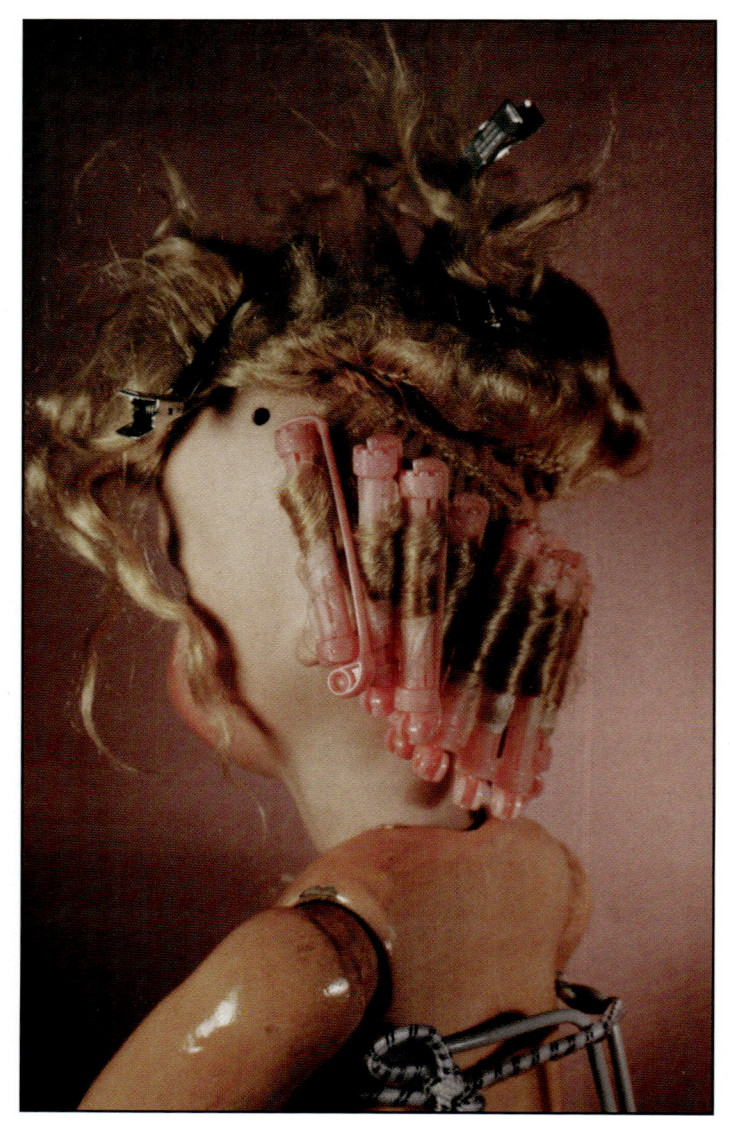

In like fashion, do curl #2 to one side of curl #1, and curl #3 on the other side (see bottom left photo on page 95).

Continue to set the rest of the bottom row, remembering to roll all curls toward the doll's face; curl direction is reversed at the center back, as you now see.

Remove the hair clips holding up the uncombed hair and section off what will be the next row of curls.

Smaller dolls (under 16 inches in height) may only be able to yield two rows of curls, while larger dolls will have more. The typical wig on a 25-inch antique bisque would probably have three to four rows of curls, but of course individual wigs will vary with the amount of hair present. Since you will probably (almost) never see a 100 year old doll with a pristine original hairdo, you need to experiment a little, and after you have set 50 or so doll wigs, you will feel much more sure about these choices!

Bottom row of curls is completed (pink curlers). (See bottom right photo on page 95 and photos above.)

The second row of curls will be placed between the curls in the lower row, and almost as low, if the length of the hair allows. What you do not want is obvious space between horizontal rows. This is one reason we start at the bottom. For this row, it is not absolutely necessary to work from the back center, as the pattern of the placement has been established by the lower row. It is a good habit to develop, however, as it will prevent many placement problems in some wigs to be discussed in later chapters. Again, remember to roll the curls toward the doll's face.

White curlers are the second row of curls (see top left photo on page 97).

By now you have set enough of the hair that you can predict how it will shape about the face. If you can see obvious problems in the arrangement, make the needed changes in placement now.

Continue with whatever remaining hair you have, taking care to maintain any original part or style of bangs that may be present. Set the third and if the doll is quite large, possibly you will have a fourth row. Again, roll the curls toward the face, placing between the rollers of the first (lowest) and second rows. If the wig is very thick, you may be literally piling the later rows over the first and second. This is a bit more difficult to do, but what a luscious set you will have, with lots of bouncy curls, when finished.

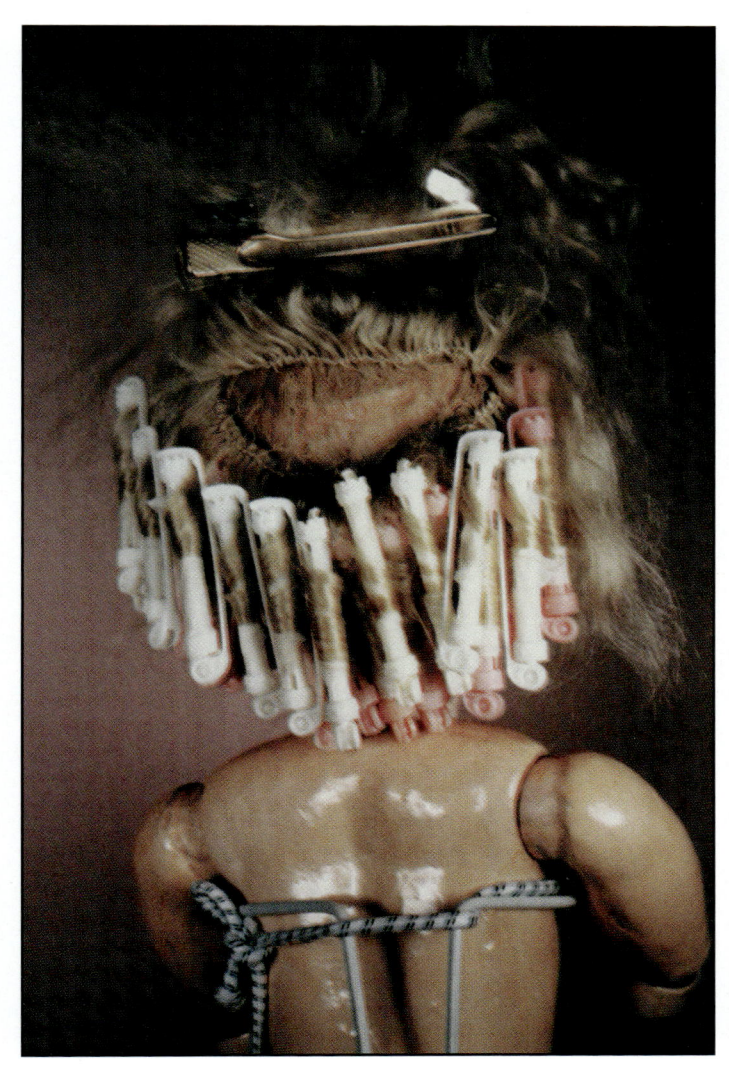

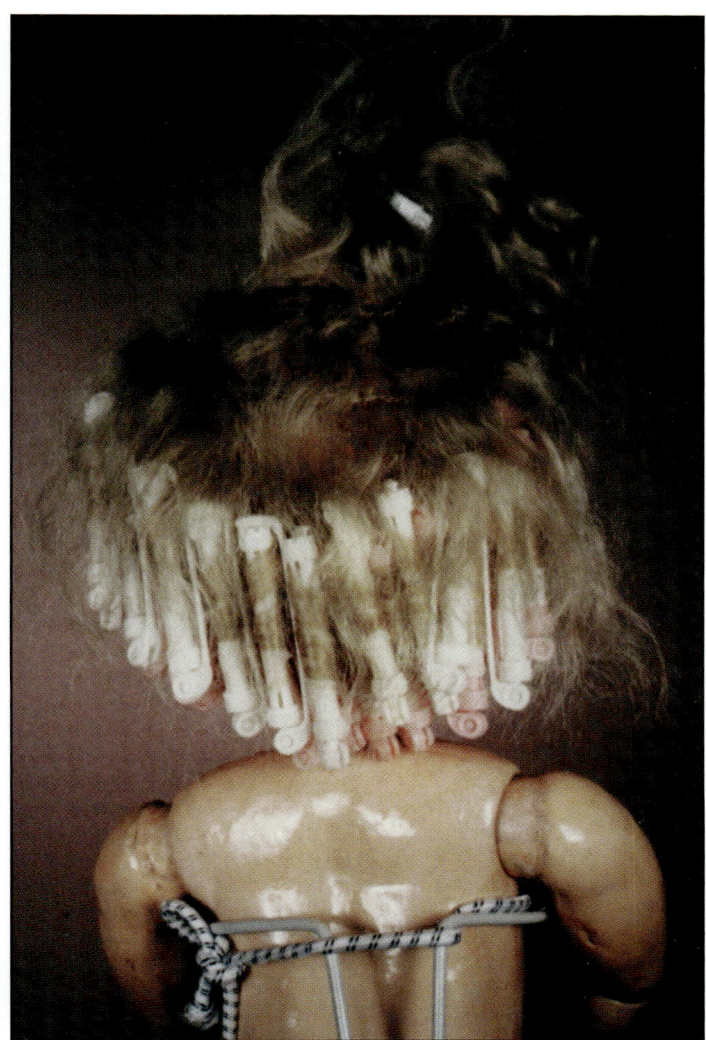

The hair for the third row of curls is brought down over the first two rows (see photos this page).

When setting is finally (whew!) completed, check all around one last time for any uneven spacing and make any corrections needed. If you feel a little more moisture is needed, spray all over lightly with water and allow a day or two for drying, depending on the thickness of the hair and the humidity of weather you are experiencing.

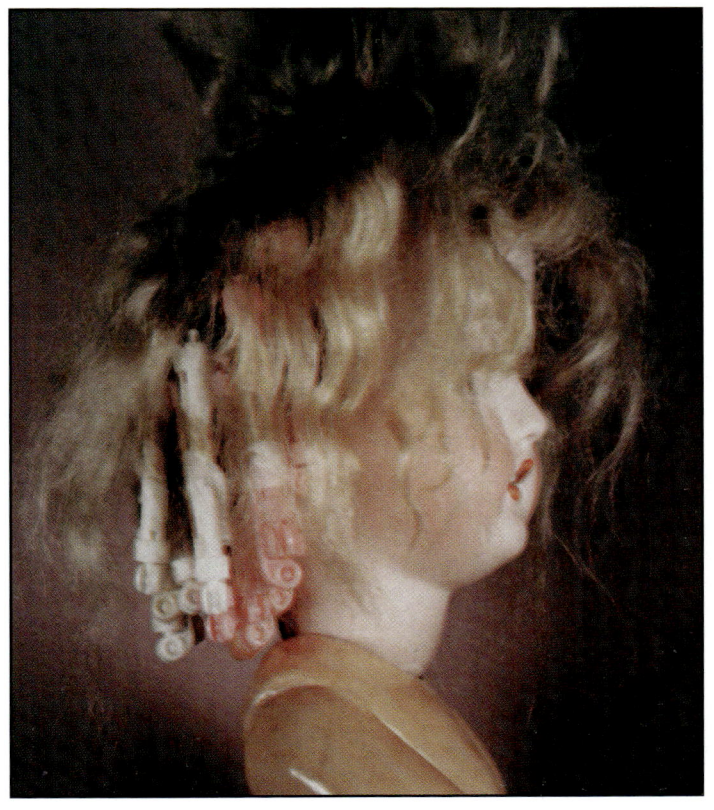

The completed set shows four rows of curlers; hair clips are used at the front center part to position the curls properly while they dry (see photos shown here and on page 99).

Resist the urge to use hair spray, or a hair dryer. The heat of a dryer may melt eye wax and cause problems you don't wish to have. Also, the temperature may not be helpful to the doll's old leather or cardboard body. So have patience, and all good things will develop; delicacy in all things is recommended with these very old children.

A Doll-Play Suggestion

We suggest that you dress your doll and do any and all manipulation, such as putting on shoes, etc., before you take down the hair. If the costuming is yet to be completed, it's a good idea to leave the hair up, so it won't be mussed when trying on clothing. If you just can't wait to see the completed hairdo, then do use a hairnet, as suggested below.

99

After a day or two of drying time, depending on the quantity of hair, remove one or two rollers carefully, by unrolling and re-rolling the curl around your finger or an object such as a pencil. Simply pulling the rollers out will leave you quite messy curls, especially with a mohair wig. Gentle handling is the key here, as is so often the case with our lovely old dolls.

Check the set. If it's a good set and the hair is dry, all is well and the rest of the rollers may be removed. If the hair is dry but has no curl....bad news: this is a synthetic wig. These will be discussed in a later section.

When all the curlers are out, resist the urge to comb the hair, except from the part down to where the curls begin.

Place a hairnet over the coiffure and pull it in at the neck so that it fits about the hairdo just snug, not tightly.

Now you may arrange curls, through the hair net, with whatever you have at hand that will function as a hair pick, such as a large hat pin or darning needle.

When you are satisfied with the styling and arrangement, spray the hair all over lightly with water. This will not cause the set to loosen much as the net is holding all in place. Rather, it relaxes the curls just a bit and lets them come together so the set looks "combed out."

Leave the net on for another day, at least, until the hair is dry. Some collectors leave the nets on indefinitely, so their dolls look as though they have just arrived from the factory! It's also a good idea for dealers, who may be thinking of selling the doll, to keep the new hairdo at its best.

Curlers are removed, the set is soft and will stay indefinitely. Front and back views of Simon and Halbig C. M. Bergmann child doll (photos shown on this page).

Wig Replacements

Often an antique doll comes to us without a wig or with a poor quality or ill-chosen replacement wig. Or perhaps the original wig is there but is so decayed there is no hope of using it. The doll owner needs to console him/herself that it's time for a replacement wig.

Wigs are sized usually in inches (occasionally mm) of head circumference; measure around the doll's head, low in back and up around ears and across the forehead. Synthetic and human hair wigs are almost always on a mesh base with some degree of stretch, so on larger heads, plus or minus an inch in size will not noticeably affect fit. Of course, on smaller heads, you will have to allow less of a margin for error in measuring. Mohair wigs usually have no "give" in the cap, or base, and when using these, size needs to be to the inch what your doll's head measures.

Many wig choices are now available, a wide variety in synthetic wigs and fewer, but very adequate choices in mohair. Regrettably, there are few options of style in human hair wigs. The best supplier in human hair doll wigs offers but one style in three colors and three sizes.

Wig Availability

Wig availability is affected by the same factors which drive the rest of our lives...economics. Manufacturers offer products which sell in volume. Volume is created in the doll supply market mainly by artisans making new porcelain dolls, whether reproduction dolls or original works by doll artists. The doll restorer who works on antique German and French dolls does not create enough volume to be able to dictate to suppliers/manufacturers, what should be produced. In other words, restorers do not have enough buying power. I once queried a supplier about a certain type of doll shoe, which they already had in production. I wanted to know if they would make a certain size in red. The reply was that if I would order ten thousand pairs, they would tool up for it. In one size, I couldn't use ten thousand pairs of red shoes in several lifetimes.

We should not lament, however, that makers of new dolls seem to be driving and thus controlling what supplies are available. Rather, I am most grateful to them, as I remember when, too recently, very little in the doll supply line was available at all. So these doll lovers have caused this industry to explode with supplies, many of which were previously not to be found. Thank you, doll makers everywhere.

This influence will also be seen in the wig styles available. Many wigs are named for a specific new artist's doll. It can be a challenge for a doll restorer to keep current with constantly new styles and names, especially when they are not involved with new dolls in their work. Again, another reason to visit your doll shop, to become acquainted with the ever-changing flood of doll supplies!

Mohair Wigs

New mohair wigs are commercially available at lower prices now than even a few years ago, in a good range of colors, styles, and sizes. New wigs can be reset just as you would an old mohair wig, given the limits of the hair's length and how the wefts were originally arranged. Of course, there is also the option of ordering custom-made mohair wigs. These are mostly done by home-based artisans and are more expensive than imported mohair wigs, but more variety is available. These wig-makers may also be willing to copy an old tattered wig. Wig-makers can be found at the large national doll shows or in the classified advertisements in doll magazines. Depending on the size of wig and whether it's handmade or commercially made (imported) you will find prices for mohair wigs range from around $30.00 to perhaps $120.00

Human Hair Wigs

Availability of human hair wigs has, in recent years, become limited. While human hair is the perfect choice for certain dolls, it does have the negative aspect of not holding a set nearly as long as mohair. Like mohair, it can be reset into almost any style, limited only by the style constructed into the wig. Curling irons may also be used on human hair wigs, but are not recommended on other wig fibers.

We are often asked to make human hair wigs from a customer's hair. Various health laws prohibit the commercial use of human hair that has not been subjected to government inspection standards. Restorers can of course use their own hair for their own personal use, but these wigs cannot then be offered commercially.

Again, sources for human hair for making wigs would be catalog companies which advertise in doll publications.

Synthetic Wigs

At the risk of being shouted out of town, or at least out of polite doll society, synthetic wigs are being included here as choices for antique dolls.

Of course, antique lovers everywhere will have a hissy-fit at the mere mention of putting these wigs on their old wonderful dolls, which deserve so much better.

My defense is simple. These are the wigs that most of our dolling friends can afford. I probably sell a hundred or more synthetic wigs for every one mohair wig that I sell, and this is not because I don't "push" mohair. My defense of that choice is that a wig is the most easily changeable doll "repair", it may be upgraded when funds allow. Many older collectors, who have watched antique doll prices rise beyond what they wish to spend, are now happily "upgrading" their collections by treating their dolls to new mohair wigs. Of course, they have to teach them to take turns, as they do these one at a time. Even dolls need to learn to wait their turn!

So price and broad choice of styles are the reasons most collectors will chose synthetic wigs, yes, even for antique dolls. Improvements in fibers in recent years have resulted in much nicer looking wigs than the synthetic wigs of the 1960s to even the late 70s. If you haven't looked at doll wigs in a while, you may be in for a pleasant surprise.

Another positive aspect of synthetic wigs is that the set will remain indefinitely, as the fibers have something we call "memory". The curling quality is set into the fiber by heat when it is spun. If these newer synthetic wigs get messy, they can be washed and reset, and will regain their curl, unlike the early synthetic wigs of the 1940s and 50s. See the section on setting synthetic wigs for all the details, and care tips.

Choosing the "Right" Wig

Your local doll shop should be able to show you a few catalogs of wig styles; there are at least four good commercial sources from which they may order for you. You may be able to order direct from one or two, but there is a great advantage in procuring your wig needs from a shop which specializes in doll customer service, and it is simply this: you need to try on a variety of wig styles and colors to make the best choice for your doll.

Just as humans find some hair styles so much more flattering than others, so will your doll! Also, there is a great aspect of personal choice and taste involved in this decision. While your local doll shop cannot possibly stock every size, color, and style or wig, they surely can show you several good styles, and then order in the color and size you need.

Nothing else will change or improve the look of your doll more than the right wig choice. Since I cannot emphasize enough, the importance of wig choice in letting your doll look her best, the following photographs will make the point.

Antique Cuno & Otto Dressel, Germany, circa 1920, wearing a new mohair wig in the shorter 1920s style. Auburn color matches brows well and brings out cheek blush. It also contrasts well with white costume.

Antique "*Viola*", Germany circa 1920, wearing synthetic blonde wig, center part, no bangs, long "sausage" curls.

A reproduction (new) *Queen Louise* wears a synthetic auburn wig, bangs and long curls look just right on this doll. Notice how well the hair shade complements the brows and cheek coloring.

A French-style mohair in dark brown proved the best choice for this reproduction Bébé Louvre. With paperweight eyes which command attention, strong style and color are needed for balance.

A precious 18-inch antique Heinrich Handwerck needs just the right wig, nothing too bold, something as delicate as she. We don't wish to overpower her sweet size and facial coloring. Who doesn't love a brown-eyed blonde? The wig is a new mohair in the traditional style of bangs and long curls.

This antique Morimura Brothers, circa 1918, needed just the right style to wear with her long white cotton nightgown. Her delicately colored brows demanded a blonde mohair, pulled up with ribbons so it won't be mussed while she sleeps.

Front and back views of this antique Kämmer and Reinhardt show a desirable doll with well executed features. She needs just the right wig to maintain her considerable value. She tries on several wigs in the photos shown on the following pages.

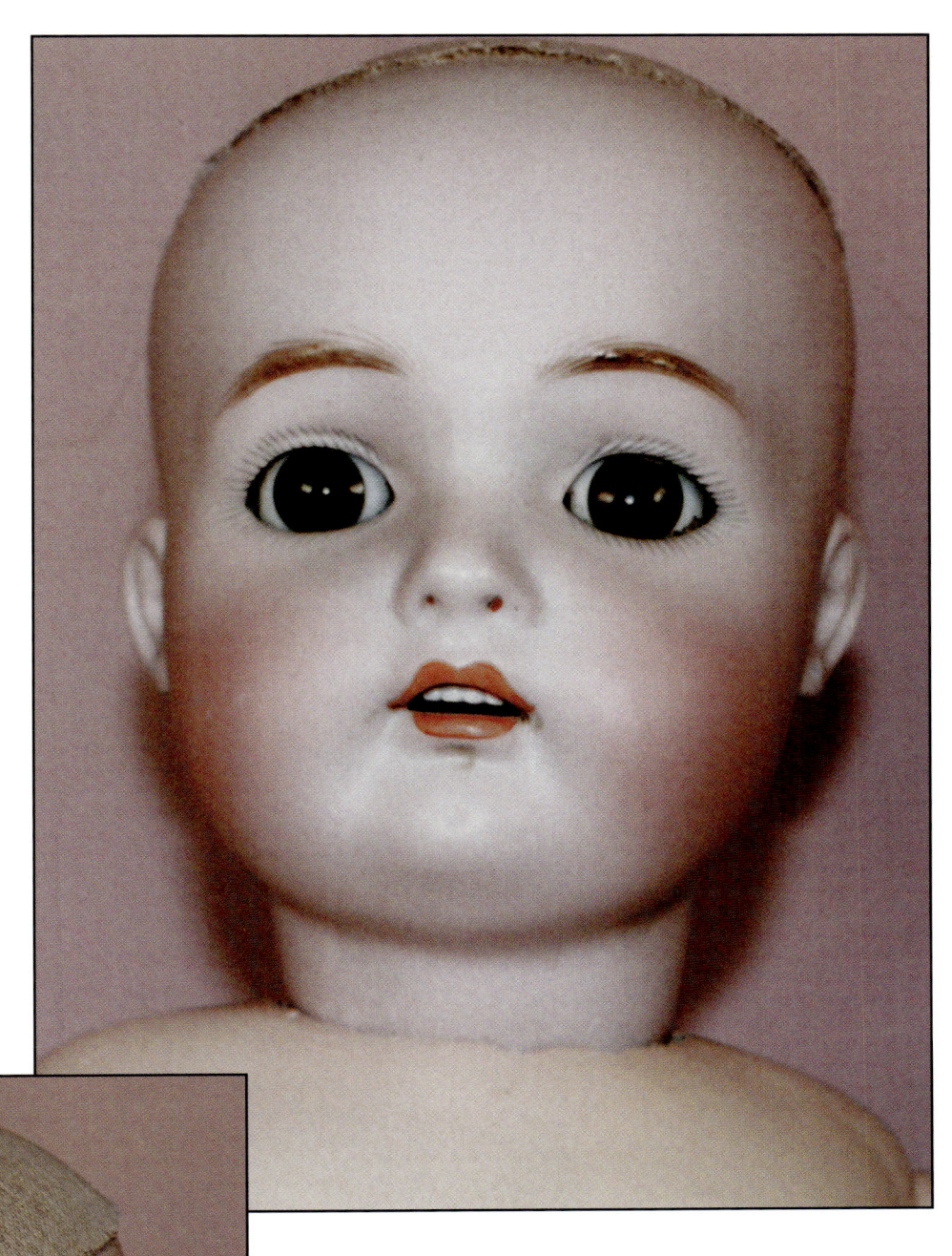

Auburn human hair wig in traditional style.

Blonde synthetic wig in traditional style.

Auburn synthetic wig in an off-center part, no bangs; large, long curls.

Dark brown synthetic wig in a long straight style, with bangs. Although this color brings out the dark brown eye color, it is rather severe with her delicate coloring. The style is not particularly flattering.

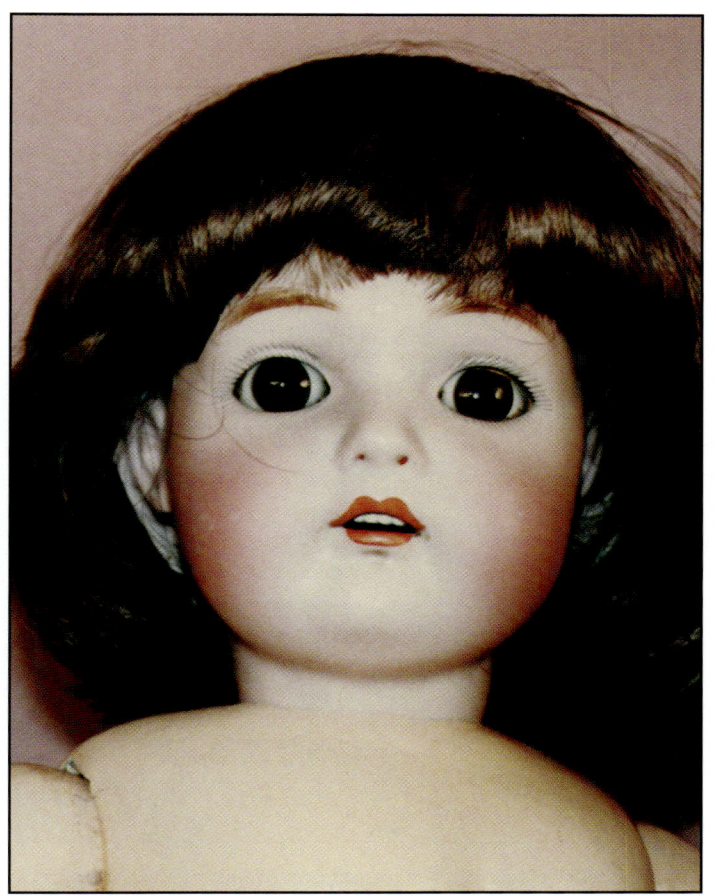

Medium brown synthetic wig in a shorter, simpler "bobbed" effect. Not a bad look, but not fabulous either. This color works well with the facial tints and features.

A slightly darker shade of auburn, in a synthetic wig, traditional style.

107

New blonde mohair wig in shorter style, spritzed with water and netted to contain its fullness. Quite charming, this wig is available in shades of auburn and brown as well.

Below: Another doll we will feature at the wig choosing party is a Japanese made Nippon, circa 1918, front and back views.

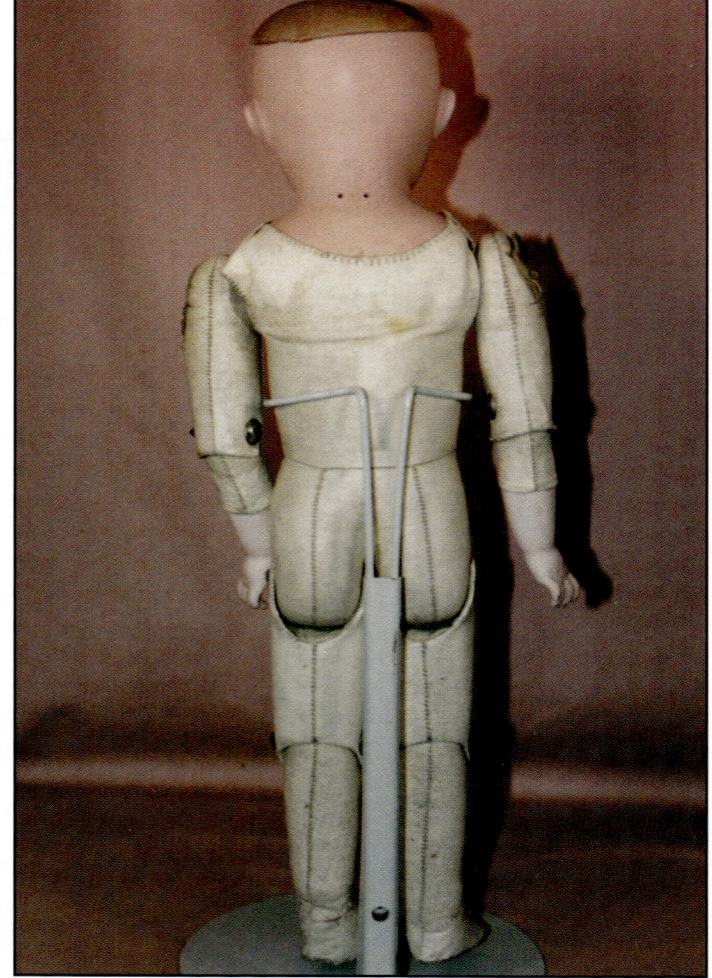

The following photos show the doll in several different wigs. She is of fairly nice quality for her type and needs the proper wig. Her features are not dramatically executed, so we want to take care not to overpower her looks with a wig that is "too much". Help us choose!

Plain center part in synthetic auburn gives her a plaintive look. *Below Left:* A simple, shorter, side-part synthetic wig in blonde, frames her face sweetly. *Right:* Lots of curls, pale blonde synthetic wig, with center part. It's cute, but a bit modern.

109

This is the same style as the wig on page 109, lower left, in medium brown. Would you have guessed?

A wonderful pale blonde mohair wig which is not wonderful on this doll. It's too "fancy" and not the best shade of blonde for her coloring.

Front and back views of a longer, fuller, center part synthetic wig in blonde. Matches brows and coloring well. It might look better if thinned a bit, by cutting out a row or two of curls carefully.

Have you decided on which wig you like for the Nippon doll? This is the type of service to ask for and enjoy at your doll shop, when your doll needs hair! Have fun with the choices!

Fun at the Doll Shop

Sometimes, a little story helps. Of course each tale features not any one real person, but rather a "combined character" as it were, of many doll lovers we have met over the years. The stories are told with affection and respect, to share the simple fun we have experienced at the doll counter. Now you won't mind if we give them all fun names?..and their dolls, too, of course.

The Wig Choosing Scenario

Case 1

Mrs. Smythe-Jones comes in with her latest acquisition from the recent doll show, and her sweet doll, little Flory Gloria (an alias), needs a wig. "Blonde," says Mrs. S.J. "I want a blonde wig. See these little wisps here? Her original wig was blonde"

And so it was. So I am listening to Mrs. Smythe-Jones, and admiring her dear new dolly, measuring the doll's head and getting out every blonde wig I have in her size. Several, to be sure. But I sneak an auburn wig into the stack. Because I have been studying her doll's coloring, especially the brows, and thinking about war-ravaged Germany when this doll was made, and speculating that the doll manufacturer would have put on the doll what was on hand in that difficult time, not necessarily what would have looked the best on the doll.

And so, after trying on all the blonde wigs and not really jumping for absolute joy at any of them, although a couple of them are better than "OK", sweet little Flory Gloria goes home cute and adorable in the auburn wig. Why? Because when it went on her head, it lit up her face and eyes and gave her a certain "personality" that the blonde choices did not. It helped her owner to see the choices of wigs in order to choose the wig that would make her doll would look her best.

Case 2

The next customer brings her little hairless, but lovely doll into the shop and asks for a new wig. "Any certain color?", I ask. "Oh, whatever you have, what do you think?" is the reply. Again I take out several wigs in her doll's size, but we do not get past the first wig. "That one's fine, I'll take that one." And so the first wig seen is glued on the doll.

Case 3

Cutting to the chase here, after several wigs are tried on the doll, the customer cannot make a choice. "What do you think?" she asks. The problem here is that her doll is so adorable and so well colored (checks, lips, etc.) that she looks wonderful in every wig we show on her.

So...the message here is that dolls are all different, and so are their owners. So, if possible, try several wigs on your doll before making a choice. Catalog shopping is a wonderful help if you do not have a local doll shop specializing in the service aspects of doll care.

If you are in a more remote area, you still have the world at your fingertips with the touch of 1-800! You can locate catalogue companies specializing in doll supplies by looking through the ads in doll magazines.

Dressing Your Dolls

Hopefully the previous chapters have dealt with all of the repair needs of your doll, and helped you get her into the best possible condition, clean, and ready to dress.

When you are satisfied with the condition of the doll itself, questions arise about the doll's clothes. If you have garments which fit her and are suitable in style, you may wish to clean or launder them, even if they don't seem terribly soiled. Many times antique dolls are acquired by purchase or inheritance which have clothes which are old, but not original, and which are sometimes quite inappropriate. This seems a proper time to address the subject of proper doll attire, and help the doll owner decide whether to use the clothing available, possibly improve upon it with added trims or features, or when necessary replace the garments entirely.

How Should the Antique Doll be Dressed?

Whether you have clothes for your doll or not, it's important to know the type of antique doll you have, and what the proper clothing style should be. Since this volume deals with antique bisque child dolls, we will address only those kinds of costumes in this chapter. If you have an antique fashion or baby doll, you will want to reference materials about those particular styles.

Antique bisque child dolls represent little children, girls usually, from the 1890s to the 1920s. While there is some latitude to be taken to accommodate personal preferences, and perhaps to use an old heirloom garment, some basic standards need to be followed. It is not always possible to know what the doll would have worn when it was made. Some later antique dolls were dressed in what could only be considered marginal attire, with little trim, poorly done seams and the like, due the economies of the times in which they were made. An example is shown on this page.

Especially for this latter reason, the doll owner may wish to imagine a nicer, but still correct costume for his or her doll. The original shift can either be used as an under or over garment, or displayed separately with the re-costumed doll.

Undergarments Should Complement the Dress Style

This reproduction Bru in the top two photographs on page 114 is shown in the preliminary stages of her make-over. Her pantaloons are put on first, then shoes and stockings in matching ecru shades. Her undergarments are in ecru cotton and cotton lace, the pantaloons gently flared, and the slip in a dropped-waist style, as her dress will be a dropped-waist style.

Chapter 7

This C. M. Bergmann wears her original costume, which is nothing more than a shift made of gauze with a little trim about the bodice. Even with the addition of shoes and stockings, she needs at least a slip to look properly dressed.

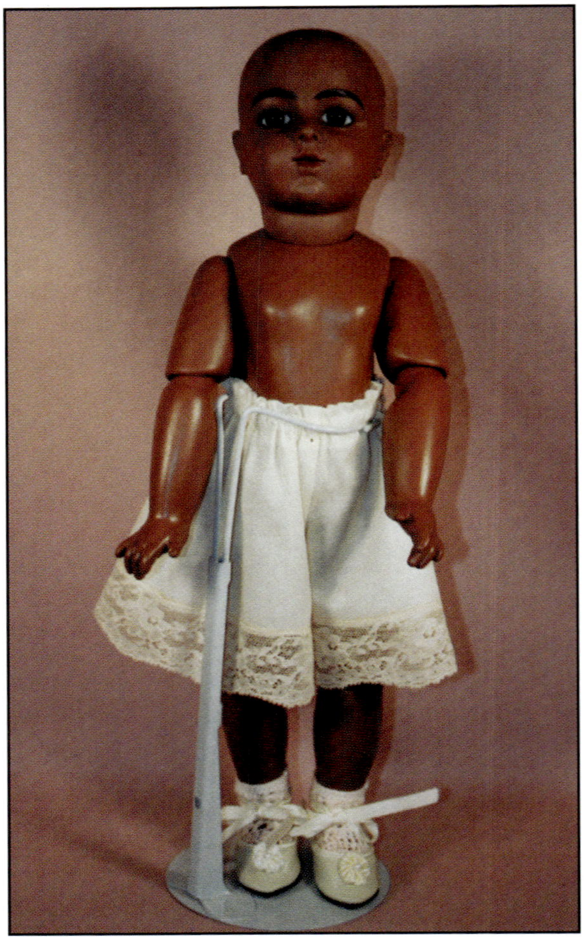

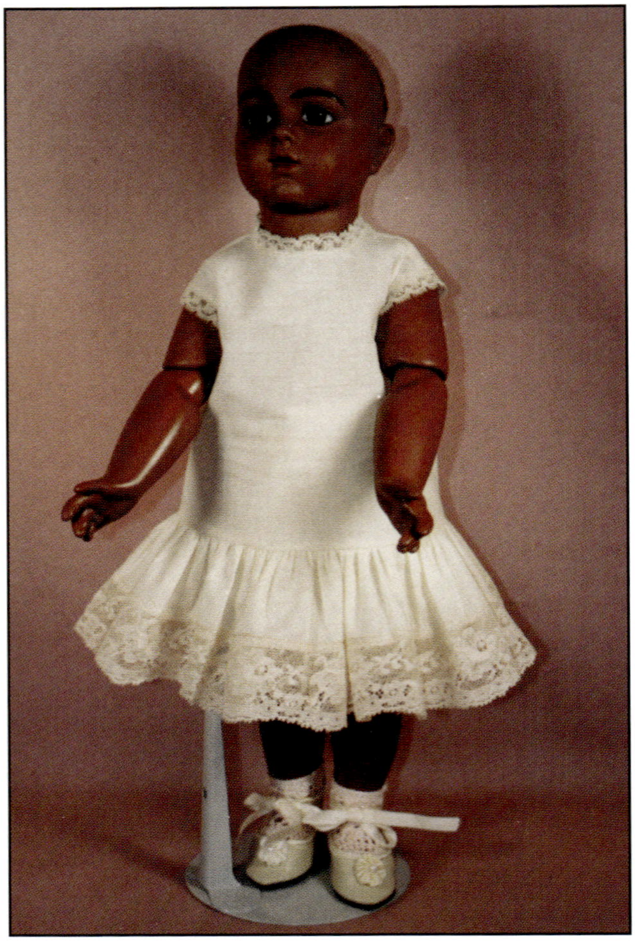

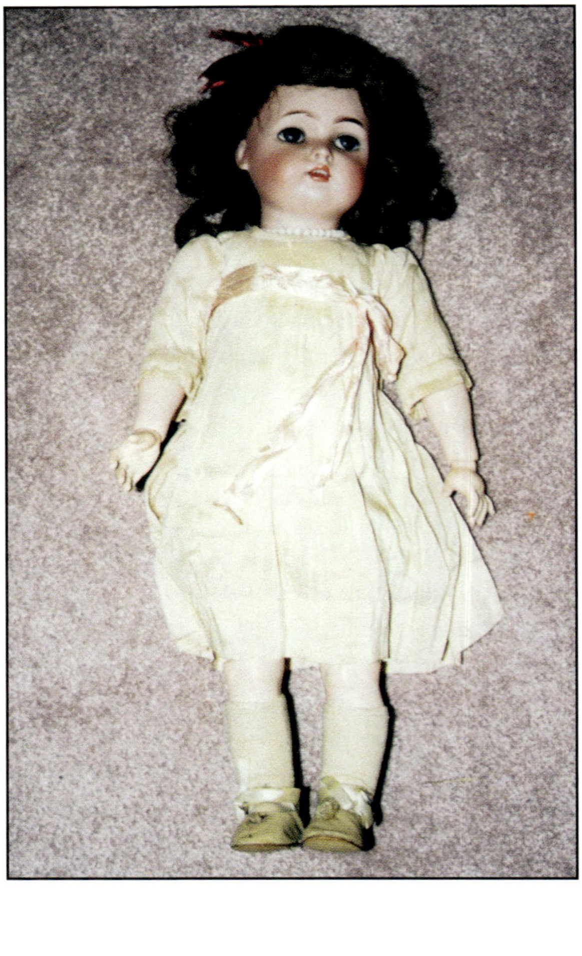

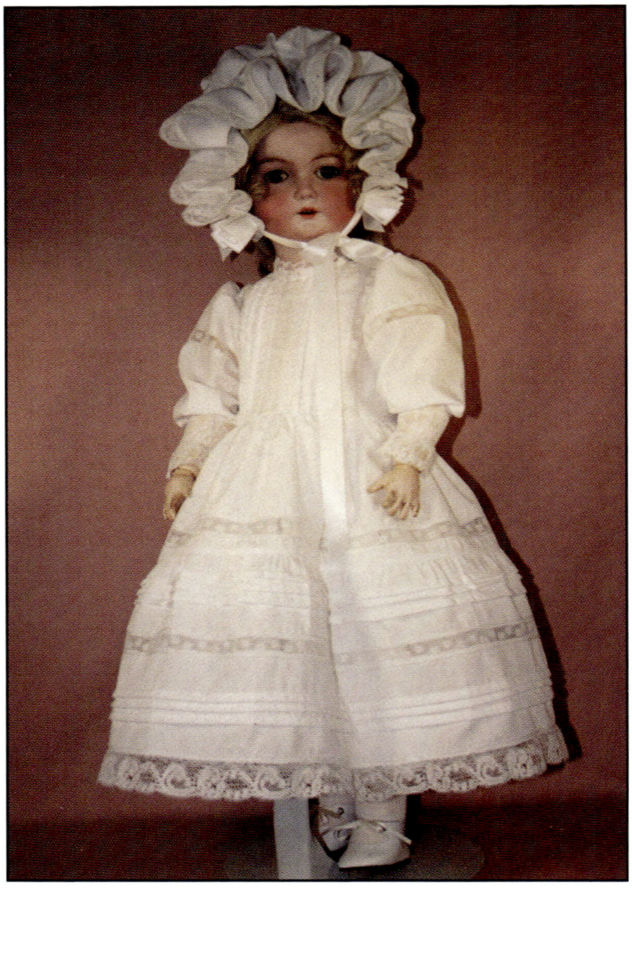

A Kämmer & Reinhardt, opposite at bottom left, in completely original condition is a wonderful find. This doll's costume will need only a light washing, and pressing, to look its best. Notice the simple design and fairly plain fabric used.

Since the costumes needed for an antique child doll will depict a typical child's wear of the period, it is simple to do a little research about children's styles, to see what the basic concepts of those styles would be.

Of course there are also many wonderful books about dolls, many of which are shown in original factory costumes, designs which one could study and emulate.

The dresses for antique German child dolls should have a skirt length not to the floor (that length would have been worn by adult women); the skirt should end mid-way between the knee and ankle. Stockings and pantaloons covered most of the rest of the leg. Appropriate shoes are needed as well. A bonnet which complements the dress is fun to add and quite correct, as well.

Typical German Child Doll Costume

Opposite at bottom right, a large C. M. Bergmann wears a white cotton dress with tucks and inset lace in the bodice and lace in gaithered tiers for the skirt, lace cuffs, and a matching bonnet. White cotton stockings and white leather shoes complete her outfit.

The doll on this page is a 32 inch antique German doll which needed a little extra detail in costuming. This doll is re-costumed in an ecru cotton moire, with tucks to the bodice, and an exaggerated bow at the hip-line. Pink ribbon under inset lace accents the dress and bonnet as well. Ecru cotton stockings and brown leather shoes complete her outfit. Isn't she a beauty?

An antique closed-mouth Kestner, mold # 128 is an especially nice doll, and deserves the extra expense of the cotton laces in her costume. The tucked bodice is cut fuller, and blouses onto a fitted waistband. The inset lace to sleeves and bodice are augmented by the wider matching lace at the hem of dress, pantaloons, and slip. She shows off her undies, as even there the lace is as lavish as on the dress. In this costume she is a lovely focal point for any room.

Dresses Re-Created in Silk and Wool
Nautical and even military themes were popular for doll clothing for antique dolls. This re-creation of a middy style is actually a skirt and blouse, done in a lightweight ivory wool, and featuring a sailor-style hat. The piping was done by machine embroidery in a wide satin stitch, which gives the sewer much latitude in choosing colors. The leather shoes were dyed to match the maroon color of the embroidery trim.

117

Silks are especially nice for doll dresses. This large Armand Marseille #390 is re-costumed in a Chinese silk print. The pattern is soft and yet vivid enough to be effective in a fairly simply styled dress. Handmade "roses" trim the ivory wool hat.

This large Kämmer & Reinhardt looks especially nice in a brown palette. The colors bring out the size and beauty of her brown glass eyes. Again a Chinese silk print is used for a fairly simple dress, with only the cotton lace edging to dress it up. Brown leather shoes, brown cotton stockings, and a brown wool hat complete the ensemble.

The Smaller Doll

Small dolls can actually be more difficult to re-costume. The use of smaller width trims is essential.

This 9 inch Simon and Halbig girl, at left, was first fitted with proper pantaloons and slip in ivory cotton, with cotton lace trims of 1/4 inch width. The same trims are used on her cranberry silk dropped waist dress and bonnet. Isn't she precious?

Antique French dolls can and often did wear styles similar to the German child dolls. Yet the French love of and flair for fashion resulted many times in doll costuming which was much more elegant than what the German dolls usually wore.

For German dolls, the use cottons or silks in whites, off whites, or soft pastels is usually recommended, for dresses which would have usually been considered "morning wear." The French dolls usually wear fancier fabrics, silks, satins, taffetas, made into dropped-waist dresses with low pleated skirts and matching jackets; hats are more "adult" looking, often with feather trims.

Typical French Child Doll Costume

The Antique Jumeau #1907 doll below wears a one piece dress and jacket combination in pink and black. Black stockings, black shoes, and a wired bonnet complete the ensemble.

A close-up, opposite page, allows you to see the dangling Austrian glass bead earrings, pleated lace trim which was used at the neck and cuffs, and of course her lovely bisque.

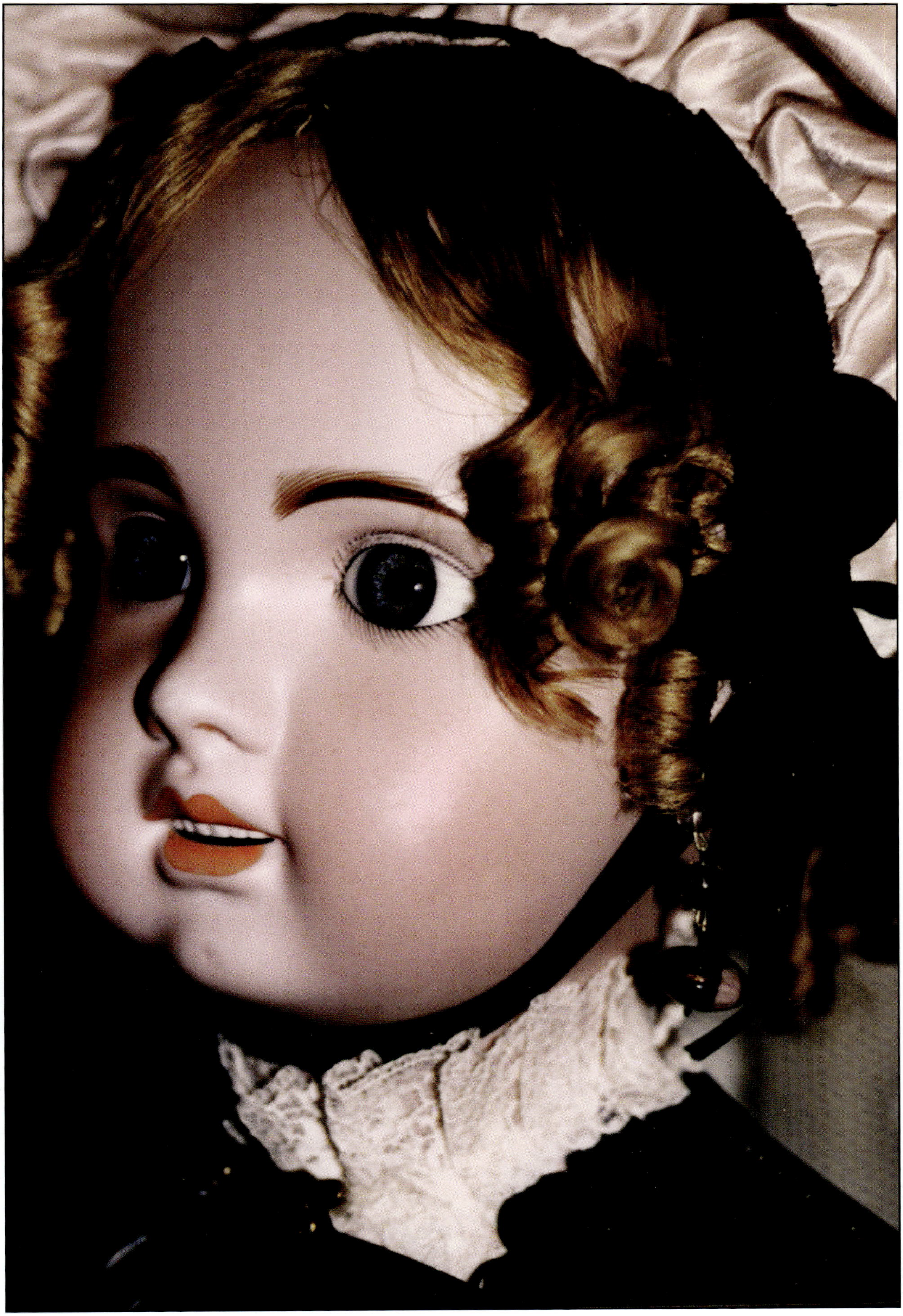

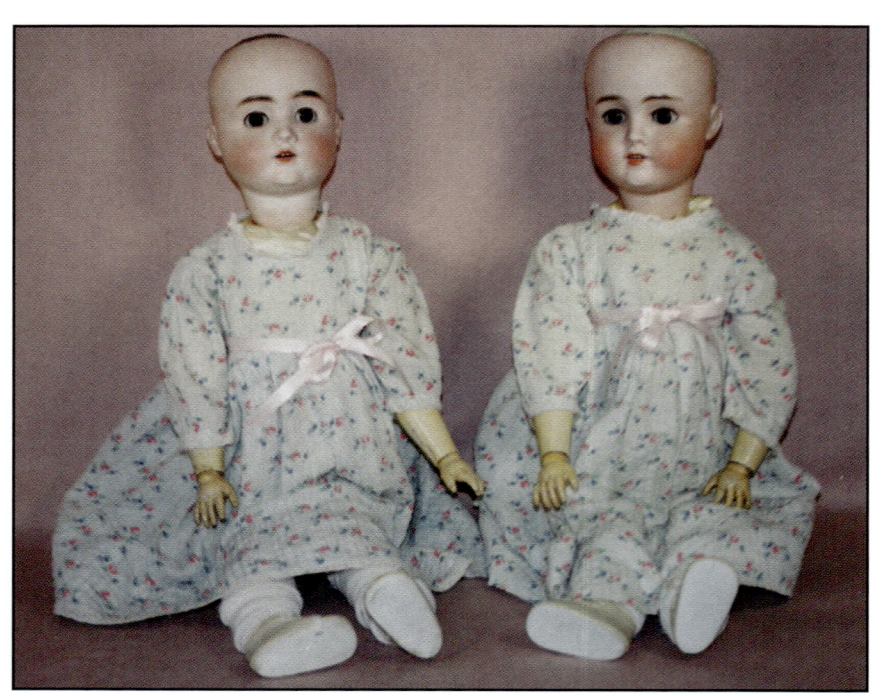

What not to "Save"

When I saw these "twin" antique German A. Wislizenus "Special dolls," shown at left, listed separately at an auction, I knew I would have to buy them both so they wouldn't be split up. Obviously they had been lovingly costumed identically by someone who cared enough to do that. Yet the dresses are of too-modern a fabric, and don't fit well enough to save. These two dolls deserve to be re-costumed.

Quite a nice Kämmer & Reinhardt child doll, is shown below, with original human hair wig and pierced ears. While her dress is made of old fabric and is an appropriate style, it is so poorly fitted about the neckline, that it cannot be altered. The sleeve length is likewise a problem. This doll also needs to be redressed. She can hand down her dress to a doll whom it will fit.

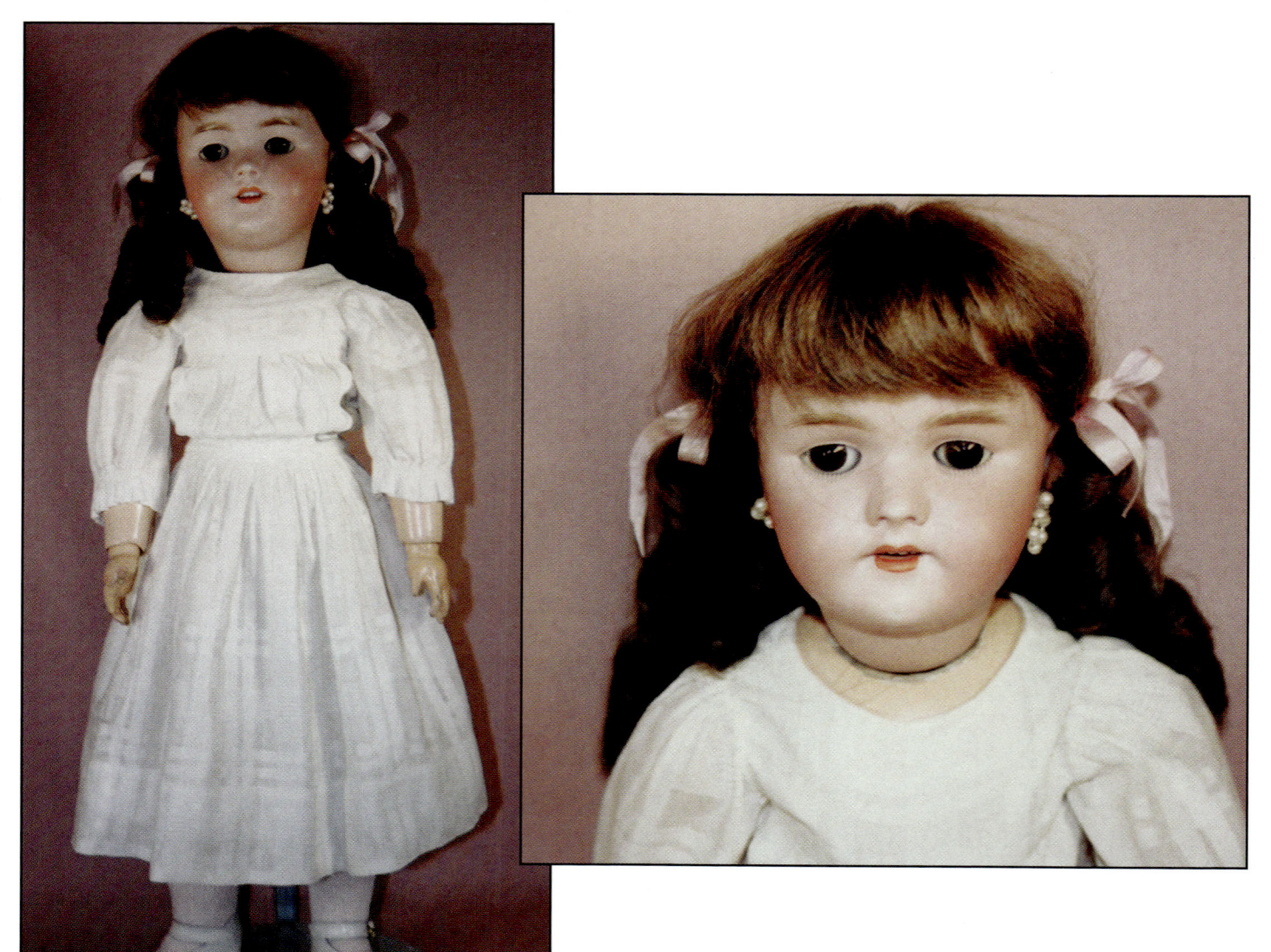

Re-Costuming with Vintage Fabrics

A Kämmer & Reinhardt *"Mein Liebling"* surely deserves just the right clothing. This lovely doll wears a "new" dress, custom fitted, using fabric pieces from three vintage garments, two ladies slips and a blouse. Her necklace is a blue topaz pendant which belonged to her original owner.

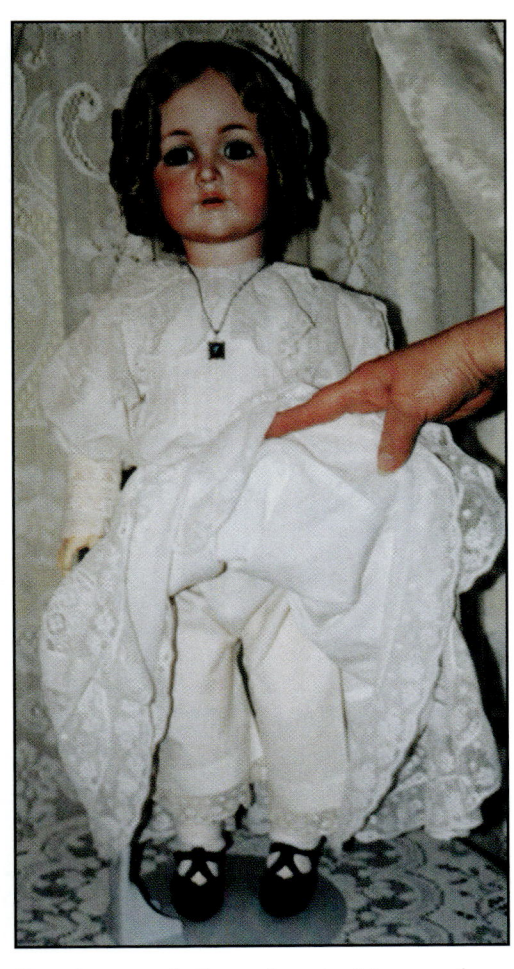

Pantaloons and slip, cotton stockings, and leather shoes are all needed.

Left: A close up of the "*Mein Liebling's*" head shows that a beautiful piece of needlework is used for a simply styled bonnet.

Below: This smaller Armand Marseille #1894 is adorable in a simple ivory cotton dress. The vintage lace trim at the hem suits her perfectly. *Collection of Patricia Obendorf.*

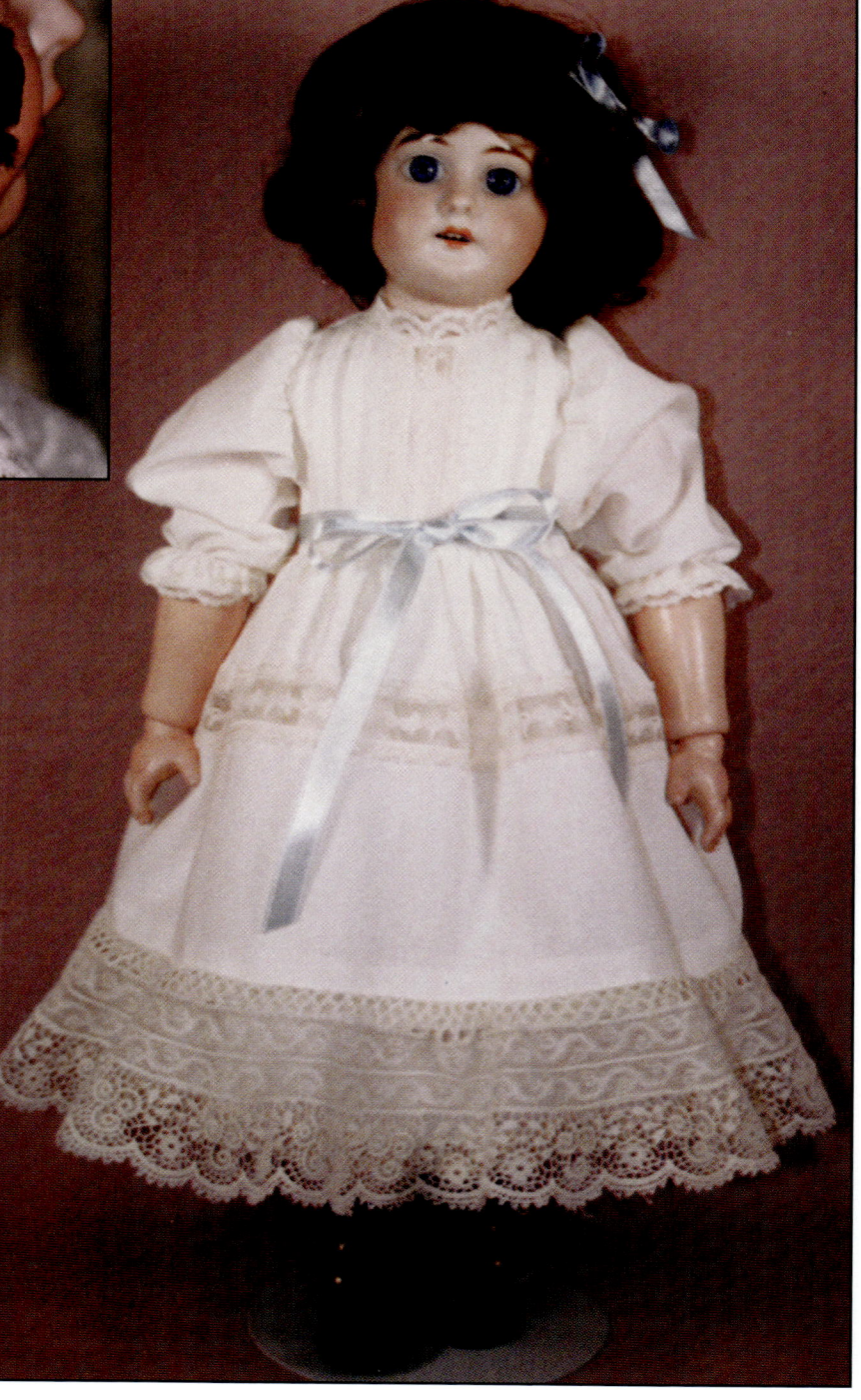

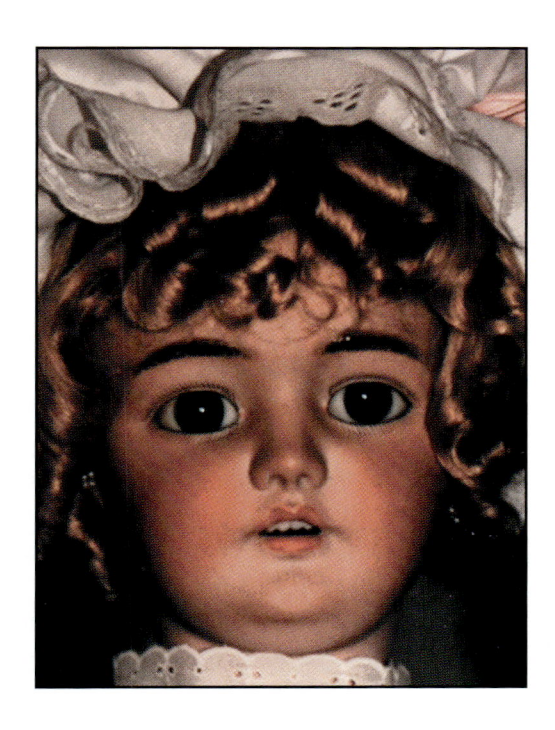

Vintage white cotton eyelet fabrics were used to create these dresses on this page.

Above, a wonderful large Simon and Halbig doll with luscious original mohair curls wears a combination high-yoke/dropped waist dress and mob hat in vintage eyelet fabrics. With large brown eyes and a beautiful facial features, she is a commanding presence.

At left, an antique Schoneau and Hoffmeister mold #1909 wears a simply styled dress made special by the use of vintage peach taffeta for the underskirt and accent bows.

125

Cleaning or Laundering Doll Clothes

If the clothes are truly antique, they will be of either cotton, wool, or silk fibers. Synthetics, of course, were not available until much later.

It is not recommended that you try to launder either wool or silk materials, these would require dry cleaning. Another problem with silks and woolens that are old, is that they will probably not be in a condition to withstand much handling, and certainly not a commercial cleaning process, so it may be a mute point. In addition, garments which are silk may also be disintegrating. So for all these reasons, garments of wool or silk may have to be used pretty much as they are.

I was thrilled to see this Armand Marseille 1894 with her original costume. The woolen dress and bonnet are too frail to dry clean. Doing a minimum of handling, the dress and bonnet were cleaned of lint, and lightly steamed to remove the major wrinkles

After 80 or 90 years, she can still wear her original woolen dress and hat. Though it has suffered some moth damage, it is quite charming and very important to her history. Note the unique style of the bonnet.

As opposed to wools and silks, cottons, on the other hand, if in reasonably good condition, and not dry-rotted, usually take a gentle hand-laundering well, and look so much the better for it.

127

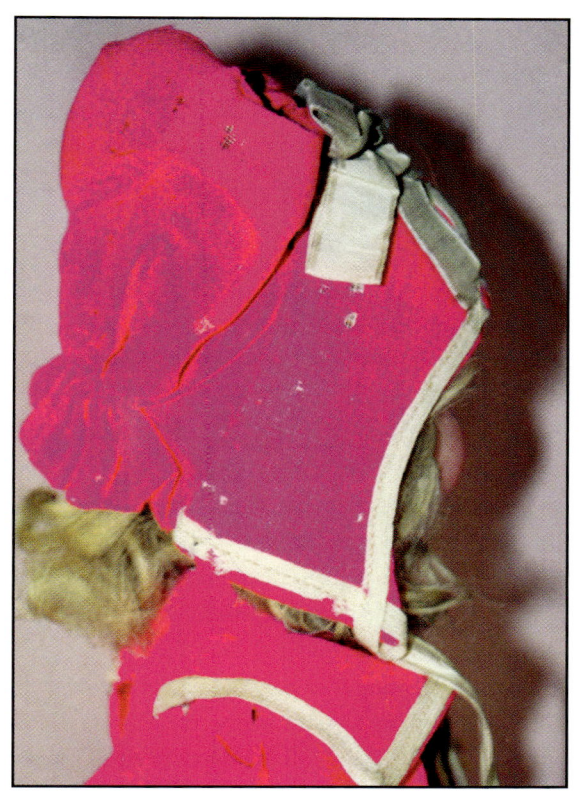

Fabrics which Require Dry Cleaning

Dry cleaning of doll clothes should be done by a knowledgeable dry-cleaner who will do them by hand, not in the cleaning machines. Such a person can be of great help in advising as to condition of fibers, type, and other problems relevant to the cleaning.

If such help is not readily available, try consulting with a textile expert, who may be located in an individual business, or at a university or museum. Of course, your public library will also have materials to aid you in this study.

Fabrics Which May be Hand Laundered

We will assume that the textiles you are handling are old, and are therefore natural fibers, that is wool, silk, cotton, or linen. It's recommended that (hand) laundry in water is attempted only on cotton or linen fibers, which are in good condition. Of course it is assumed that we wouldn't even consider putting doll clothes into a washer!

When deciding if or how to launder cotton or linen, you will need to know how to identify dry rot. This condition is caused by a small organism, yes, a little living creature, which goes through the molecules of cellulose (cotton and linen are plant fibers), eating it away. What remains is the "shell" of the cellulose, which makes the fabric feel brittle, tear at the slightest touch, and tend to disintegrate badly, when put into water. If garments are dry-rotted, they will not survive even the gentlest handling or laundering, and so, unfortunately, must be left as they are, and probably must be replaced, so the doll may be displayed in a well dressed state.

Supplies Needed for Hand Laundry of Doll Garments

Most of the items needed are common household goods which you probably have
on hand.

Equipment includes:

- White vinyl wash basins (clothes may need to be soaked for days and you won't want to tie up a sink!)
- Large White plastic spoon, check that the surfaces are smooth and won't snag garments.
- Gloves
- Timer, to time use of bleach
- Old towels to protect counter tops, and to lay out garments to dry.
- Small scissors, needle, thread, snaps
- Lint roller
- Detergents and Cleaning Aids include:
- A good laundry detergent such as ERA
- An oxygen "bleach" such as Clorox-2, for whitening and brightening
- A chlorine bleach, such as Clorox (do use carefully) necessary to remove stains from cigarette smoking, mildew, molds, and fungi.
- Dishwasher detergent, crystals or gel, such as Cascade or Sunlight
- Spray sizing and/or starch
- A good steam iron

Time and Patience in abundance!

Preparing Garments for Hand Laundry

Preparing a garment for laundry may take longer than the actual washing! These points are important to consider for successful washing of delicate old garments. Remove anything which may run or stain

Before hand-laundering, remove anything from the garment which may run, or cause a stain. This would include straight pins, safety pins jewelry, even snaps, which may cause a nasty rust spot. Look especially for straight or safety pins, as these were commonly used on doll clothes. Anything metal needs to be removed.

Bear in mind that many materials that were used for trims on doll clothing were those of lower quality, therefore cheaper. These would include some ribbons, and little paper flowers, feathers, a whole list of items. Use your judgment, and err on the side of caution. Many of these trim items cannot be laundered successfully, dyes in them will usually run, and can ruin the entire garment. So you will need to remove these types of things from the garment before it goes into water. Sometimes trims were stapled on instead of being sewed, and these tiny staples also need to come off the garment, as they may cause rust spots.

Rotted elastic is another item to remove before laundry. It is sometimes found in sleeves or bloomers, all dried up and stretched out. There is no point in laundering it, as it causes the drying time to be longer. Replacing it would be done when the item is clean, dry, and pressed out.

Basic Laundry Methods for Doll Clothes

Forgive the repetitions you will find in this section. I would rather mention some items again, in case you do not read each section. Assuming your cottons or linens are in good condition, launder as follows.

- (Separate whites from colors.
- Launder colors only with like colors.
- Remember that many fabrics are weaker when wet. Especially when handling old garments that are wet, support the item from underneath when moving it about. This is also why it is not recommended that you hang these items to dry. It's best to dry them laid out on old toweling.

• Use a white (very important) plastic or vinyl dishpan. In a white pan, you will be able to tell by the color of the water how the removal of soil is coming along, when the garments are clean, or when colors begin to "run." Using a dishpan of colored plastic will often not let you really see what is happening in the water. It's necessary to have a pan large enough to accommodate the size of the items, and allow the water to move freely about it.

• Use warm, not hot water.

If whites are very badly soiled, due to coal heating systems in old homes, or exposure to years and years of cigarette smoking in the home, more drastic cleaning measures are needed. A teaspoon or two of dishwasher crystals, thoroughly dissolved in a gallon of warm water, is used for the first couple of soakings, until these heavy stains are removed. Using the plastic spoon, stir and gently agitate the garments often; if the water is quite soiled right away, change it, and proceed. There is no point in soaking the garments in very soiled water. Then follow with the regular detergent soakings.

For other problem stains, such as mildew, mold, and the tars from cigarette smoke (these can be the most difficult to remove) you may need to use a chlorine bleach. Caution is urged, as bleach can be very damaging to textiles. Use only a capful, that is 1 to 2 teaspoons, per gallon of warm water. Set a timer, for a short period, say five minutes, and check the progress. Continually set the timer so the garment is not left in the bleach solution too long. A maximum of 20 minutes total should be observed, followed by ten rinses, to assure that all the bleach is rinsed out. If you use chlorine bleach on a very old frail cotton, you should realize that you may be risking the garment. Likewise, using it on colored fabrics may remove the color. Testing a small spot at the back of the garment is advised.

For garments not overly soiled, (or following the previous step, for very soiled garments), use a teaspoon or two of laundry detergent per gallon of warm water. Soak at least twenty minutes, changing the water whenever it begins to look dirty. When you are satisfied that the garment is clean, begin rinsing in warm water, for twenty minute periods, and do at least four, but preferably ten rinses. This seemingly extreme number of rinses is needed to really remove all of the cleaning products, especially if the garment has heavy gathers, ruffles, or layering of fabric.

When all of the rinses are completed, you can gently squeeze the water out of the garment by hand, but do not twist or wring it. Rolling it up in an old terry towel is best, to remove most of the excess moisture.

Then lay it out on a dry towel; finger press as much as you can around collars and other areas which may be difficult to iron.

Our sample for this laundry lesson is a dress belonging to a pretty doll aptly named "Princess", above. When adopted, Princess wore an old, charming, but quite soiled pink floral dress. Her dress and undergarments suit her personality and are quite cute, so we want to retain them for her, but certainly they need to be laundered.

129

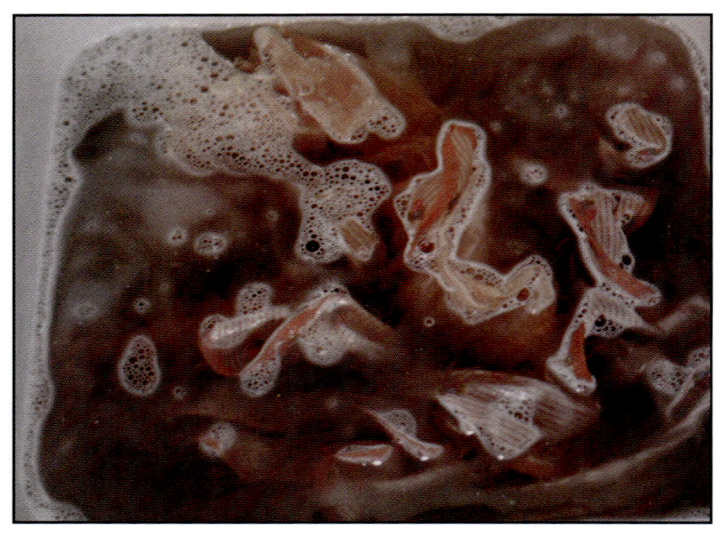

Princess's dress in its first soak. This photo shows the importance of using a white, rather than a colored, basin for the laundering process. The degree of soil in the water is easily visible, as would be any running of the dyes.

After several soaks in detergent, the water remains clean, which indicates that the dress is ready for rinsing.

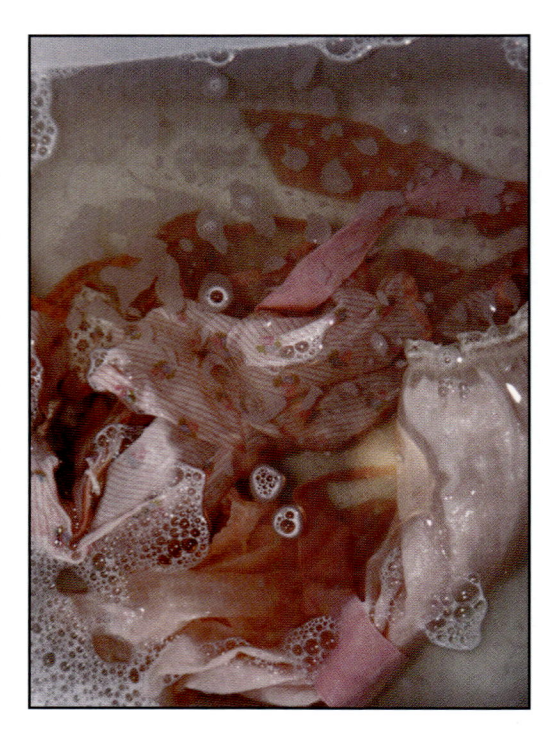

Her garments are clean and pressed. Princess has had her hair done; we have given her appropriate stockings and shoes, and she is ready for display.

Pressing Doll Clothes

With their tiny bodices and sleeves, doll clothes can be difficult to press. The use of a sewing "ham" is recommended, if a small size is available. If a ham isn't available, you can fashion one by rolling up a small hand towel, preferably one that is smooth terry, and has no raised pattern.

The second consideration when ironing, is temperature. Try a lower setting than you would normally use for the fabric in question, and then increase it as needed. It must be stressed, when ironing old garments, to be gentle with these old textiles. It's all too easy to poke a hole right through a sheer cotton with the iron.

So, carefully proceed as follows:

If there are repairs needed on the garment, press those areas first, lightly, and proceed to the repairs. Handling will undoubtedly cause some rumpling, so there's no point in pressing to perfection until the repairs are completed.

Spray garment lightly with spray sizing or starch. Using the rolled terrycloth inside the garment, steam-press the bodice and neck areas first.

Stuff small bodices/puffy sleeves, with cottonballs or paper towels to hold shape, and steam.

Iron the skirt last, using temperature as low as needed; use a little more spray sizing or light spray starch, if needed to finish the ironing.

A little white vinegar on a pressing cloth is useful to set a sharp crease, such as for pleats (or remove a pesky wrinkle).

Hang or lay out the pressed garments. Make sure they are really dry, before dressing your doll.

Schoenhut Baby and garments, before laundering.

Schoenhut Baby, garments laundered and pressed. Isn't this an improvement!

Antique china head doll, with rather soiled garments, luckily all cotton, so they are suitable for laundry. This could be considered a dirt emergency.

Some tedium was involved in removing these tiny garments from the doll; the extent of the soil is now more apparent. We are dealing with heavy soot from a coal or oil heating system, as well as a rust fungus on one of the undergarments.

After removing the beaded collar from the dress, it is, soaked separately, shows heavy soil in the water immediately. Several soakings over several days will be needed.

The beaded collar has been repaired and re-attached to the bodice of the dress. Cleaned and pressed, our little lady is ready for another hundred years.

White (we are being optimistic here, we hope they will be white!) undergarments are soaked separately, using dishwasher crystals first, due to heavy soot. Following several rinses, chlorine bleach was necessary due to "rust" fungus on slip.

When received, this little Armand Marseille doll charmed everyone, even though soiled and disheveled.

It is so wonderful to find an antique doll with such a cute, original costume! In addition to the laundering of her dress, she needs a little help with her hair, and of course some shoes and socks too! We will have to watch carefully for signs of the red dyes running, when we launder this garment.

At left, a final laundry soak of the vintage dress shows the water remaining clear. Fortunately the reds in this garment were colorfast!

These lovely girls (left, antique *Pansy*, right, antique Heinrich Handwerck) luxuriate amid an assortment of new silk and cotton doll dresses. What to choose, what to choose!

When restoring a doll costume, little things really mean a lot. These are little touches which are noticeable not when you do them, but rather when you don't! In this case, we had four small things which needed attention, in addition to the laundering of the clothes.

Simply listed, they were:
- Shoes and stockings!
- It's a bad hair day!
- Belt needed work!
- Two buttons missing from collar of dress.

The ribbon used for the original dress belt was quite tattered, and could not be salvaged. We replaced this belt fabric, using a fabric of similar color. It was bonded to a head-bond backing and cut to the proper size. With the original buckle reattached, it looks great.

Another little repair that was needed, was the re-creation of two buttons, which were missing from the collar. Again, it was easily done, as the buttons were actually embroidery stitches over tiny cardboard discs. They were easy to duplicate.

Before dressing, we styled the doll's hair, and put a hairnet over it, to protect it during the dressing process. If the wig had been off, we would have left it until last.

So, as the saying goes, little things mean a lot, and they certainly did to this little doll.

New belt, two new buttons, shoes and stockings, and a little straw hat! She looks like a new girl!

New Doll Clothes

If your doll has nothing to wear, don't despair! It's possible to find new doll clothes which look old and are just perfect for your doll.

These and similar doll clothes are usually available at the larger doll shows, and through doll magazine ads. Be sure to take your doll with you for a fitting! (They love to go shopping.)

In addition, you may also consider making a doll costume yourself. Patterns are available through doll specialty sources, and sewing for your own doll can be a great joy. You will also have the advantage of getting exactly what you have in mind.

Enjoy Your Time with Dolls

I hope this volume on doll conservation has been of help in de-mystifying the many kinds of "flaws" we often see on dolls.

One of my goals is that this book would help you to more critically observe dolls, and more easily identify those conditions which aren't too difficult to correct, and also those conditions which are not possible to remedy.

Possibly you may also now consider buying a doll which you would not have considered earlier, thinking the repairs or restoration were too complex.

So do enjoy your dolls, and all your hours with them.

Another sample of dolly laundry follows.

Above: Rather soiled white cotton dress and slip, of Victorian period. Minor mending to dress was done prior to laundering it.

Left: Laundered and pressed, her whites are now successfully white!

Below: The first detergent soak shows quite a lot of soil going into the water immediately. Many soaks were needed to clean these garments.

All My Doll Days

I was raised on a farm in Illinois, the middle child of seven, with six brothers! While I reveled in doing all kinds of "boy" things with my brothers, when they would tolerate it, I always enjoyed hanging at my mother's elbow while she cooked, canned, washed and ironed, and supervised the lot of us. She would assign me chores, but rarely told me how to do anything. Rather she let me find my own method, either copying my observations of how she did things, or inventing what I thought was a better way. There was always a pile of family mending, and I learned to use a sewing machine at an early age. I put patches on untold numbers of what are now called "jeans," but what we farm kids knew as "overalls."

While we lacked air-conditioning and a television, and many other unknown amenities today's children require for survival, we had a rich life in terms of opportunities and experiences. We were always on our own for entertainment, when our chores were finished, and we found amazing pastimes for our precious play hours.

On hot summer afternoons, I'd often spread a blanket under a large shade tree, (it would be cooler outdoors than

Mary Caruso, age 7, with her most-loved doll.

inside), and play with my dolls and two puppies, probably interchangeably! I do recall that the puppies took many rides in my doll buggy. The wicker buggy was a gracious and most welcome hand-me down from the town's bank owner, a most lovely gentleman.

I would often implore our father, who ruled our lives firmly, to tell my younger brother to play dolls with me. If dad said it, that was it, and poor Jim spent, I'm sure, a miserable afternoon helping me make a doll dress from one of dad's old shirts, when he could have been catching frogs or some such thing, with his twin brother.

It is a warm memory for me, even now, the time I spent on an old treadle Singer on our back porch, sewing up yet another garment for my favorite doll, a little rubber drink and wet baby doll of the late 1940s.

My mother loved to have a small flower garden, in addition to the huge vegetable garden she always planted to help feed the nine of us.

She once showed me how, if I pulled the stamens from a particular flower, the blossoms could be fitted, one into the next, to make a necklace. This intrigued me, and I would often make one of the flower chains into a crown for my dolls.

Such were the joys of childhood, that they have remained with me, at times even sustained me. The simple love of dolls also provided an instant common ground of friendship with other doll lovers, no matter where I found myself.

It is this joy that I wish for each of you, summarized in the following poem.

The Dolls' House
Mary Caruso
1996

Especially on a rainy day,
I climb the attic stairs,
up the fifteen steps, I'll be
in childhood fantasy.

Over in the corner where
my childhood things are stored,
I pull back the dust covering
to open my toy cupboard.

Now my diary pages fly
back to a simpler time,
when I never minded rainy days,
I'd make the upstair climb.

It's the House of the Two Sisters
I named them Vi and Nell,
Vi of course is Violet,
and Nell is Darling Nell.

The Sisters come into their parlour,
and ready things, with me
tiny little cups and cakes,
imaginary tea.

Then I take them to the porch,
They tend the flower beddings,
and check the postal box to see
if there are penpal letters.

When evening comes they go inside,
and after simple supper,
Vi's turn to do the dishes,
and Nell's turn at the spinet.

It's such a child's fond joy to me,
that they get on so well,
they trade a careful good-night kiss,
between their china faces.

I close them safely back inside
the dollhouse where they dwell,
safely waiting for me there,
when rainclouds come again.

Enjoy Your Time with Dolls

I hope this volume on doll conservation has been of help in demystifying the many kinds of "flaws" we often see on dolls.

One of my goals is that this book would help you to more critically observe dolls, and more easily identify those conditions which aren't too difficult to correct, and also those conditions which are not possible to remedy.

Possibly you may also now consider buying a doll which you would not have considered earlier, thinking the repairs or restoration were too complex.

So do enjoy your dolls, and all your hours with them.

INDEX

A

Acrylic Eyes 12, 66
Armand Marseille, photographs 13,
..........23, 63-64, 70-73, 118,
................124, 126-127, 134
Arms, Sewn-On 52-54
Arms, Strung.............42, 43, 54
Arms, Wired-On 52
Attachment of Bisque
 Socket Head 44, 49

B

B-62 Doll, photograph 17
Ball Joints 29-43, 46
Ball Joint Repair 29-39
Bergmann, C.M., photo-
 graphs 14-113
Bisque 5, 7-8, 10-11, 14, 26-28
Bloomers..............................128
Bodies, Ball-Jointed 16, 29-43
Bodies, Cloth 20
Bodies, Composition 30-31
Bodies, Crude
 Five-Piece 7, 45-46
Bodies, "Put-Together" 17
Bodies, Repainted 18
Body Authenticity 16
Body Braces 40
Body, French Style 16
Body, German Style 16
Body Types 16
Bye-lo, photograph 11

C

China head, Photo-
 graph 7-8, 132-133

Cleaning of Bisque 26-29
 China 26-28
 Composition Bodies 30
 Human Hair Wigs 89-91
  Leather Bodies 51
 Mohair Wigs 89-91
 Oil Cloth Bodies.................51
 Synthetic Wigs 89-91
Cleaning Products and
 Supplies List 25, 30-31, 39
C. M. Bergmann, Photo-
 graphs 14, 113
Costuming 113-136
Curlers 93-99

D

Doll Bodies ..7, 16-18, 20, 29-43
Doll Stories 10
Dolls, Antique 12-21
Dolls, Antique Reproduction ..11-12
Dolls, "Put-Together" 17

E

Elastic Loops 39, 49-50
Eye Quality 13
Eye Repair 65-88
Eye Repair Materials List........67
Eye Rocker 88
Eye Setting, Fixed Eyes 65-66,
 ....70, 74
Eye Setting, Sleep Eyes 66,
75-83
Eye Sizing 65
Eye Type 65
Eyes, Glass Paperweight 73-74
Eyes, Misaligned 76-77

Eyelid Painting 68-69
Eye Wax 71, 74

F

Finger Lining 38-39
Finger Repair 37-39
Fixed Eyes 65, 70-74
Floradora, Photographs 70-73

G

Glass Eyes 5-6, 8-9, 12-13,
 ....18, 65-66, 68, 77, 81,
 ....83-85, 87-88

H

Hair, Human 101
Hair, Mohair 101
Hair, Synthetic 101
Hairnets 92-93, 108
Halbig, Simon &, Photo-
 graphs 94-100
Handling of Dolls 21-22
Handwerck, Heinrich,
 Photographs 7, 104, 135
Hooks...............................39-50

J

Jumeau, Photographs......120-121

K

Kammer & Reinhardt, Photographs............6, 15, 23, 49-51,
 ....105-108, 114, 119, 122-125

INDEX

Kestner, Photographs.............5-6
Kid Leather Bodies 19-20
Kid-o-line Bodies 20

L

Laundry of Doll Clothing 126-136
Leather Body Repair 51-58
Leather Tears 55-58
Legs, Composition,
   Resetting 34-35
Legs, Wired-On 54-55
Limbs, "Staked" 45

M

Marseille, Armand: Photographs 13, 23, 63-64, 70-73,
   ....118, 124, 126-127, 134
Mein Liebling, Photo-
   graphs 123-124
Morimura Bros. Child Doll,
   Photographs 16, 18

N

Neck Button........................43-44
Neck Hook....................43-45, 49
Neck Socket........................36-37

O

Oilcloth Body 18, 51
Open Mouth 28

P

Painting of Repaired Areas......31
Pansy, Photographs..........16, 135
Paperweight Eyes 65, 73-74
Parian 5, 7, 10
Parian Doll, Photograph 7
Plaster Eye Pivots 68, 75, 77,
   ....80, 82, 88
Princess, photographs 129-130
Porcelain 5, 10-11, 16, 29

Q

Queen Louise Doll,
   Photographs 13

R

Reinhard (Kammer and...)
   Photographs ..6, 15, 23, 49-51,
   ....105-108, 114, 119, 122-125
Repair, Leather 51-58
Repair, Oilcloth......................51
Repair Products and
   Supplies........................30-31
Replacement Body Parts..........31
Replacement Cloth Body....62-64
Reproductions 11-13 103, 114
Rocker, Eye 66-68, 71, 75-76,
   ....80-81, 83-86, 88

S

"S" Hooks 39, 41-42, 46-49
Schoenhut Baby Photographs ..131
Setting Wigs.....................94-100
Shampooing Wigs....................90
Shoulderplate Heads 62-64
Sleep Eyes,
   Setting/Installing 75-83
Sockets, Body 32-37
Sockets, Neck 36-37
Socket Repair.....................32-37
Socket Replacement 3
Sticky Wax 29, 65, 71, 74
Stringing Cord........................39
Stringing of Body, Diagram 41
Stringing Supplies...................39
Stringing Techniques 39
Synthetic Wigs 89

T

Teeth, Plastic 28-29
Teeth, Porcelain 29
Teeth, Replacing 29
Tongues 28-29
Torso, Cloth, Replacement ..59-62
Torso Lining 59-60

U

Upholstery of Damaged
   Body 57-58

V

Varnish 31

W

Wash, Paint, for Color Match....31
Wax, Sticky............29, 65, 71, 74
Waxed Eyelids, Cleaning 67
Waxed Eyelids, Repainting ..68-70
Wig Care 89
Wig Choices 112
Wig Fibers 93-94
Wig, How to Glue On 91-92
Wig Mending 92
Wig Shampooing 89
Wig Setting Process..........94-100
Wig Setting Techniques 93
Wig Styles 101-111
Wire Loops 46
Wired-On Arms/Legs 52, 54-55
Wilizenus Photograph............122

Biographical Information

Education: Bachelor of Science, Mundelein College, Chicago, IL

Positions Held:

    Medical Research, Loyola University, 3 years

    Teacher, Jr. High Math and Science, Chicago Diocese, 10 years

Awards:

    Illinois Junior Academy of Science Chairman's Award, 1997

Memberships:

    past member, Illinois Science Teachers Association

    past member, Minnesota Quilters Association

    United Federation of Doll Clubs

    Madame Alexander Collectors Club

    Doll Artisan Guild

Activities:

    Diocesean Science Curriculum Committee Member

    Science Teachers' Newsletter

    Science Teacher In-Service Workshop Presenter

    Volunteer Math Tutor

    Volunteer Akron Historical Society

    Co-Founder/Chairman, Akron Doll Show

Owner, Doll Works, Doll Shop and Hospital, Stow, OH

Hobbies: Travel, Antiquing, Reading, Gardening, Old House Restoration, Study of Anthropology